Introduction to farming systems

MICHAEL HAINES

Professor of Agricultural and Food Marketing,
Department of Agricultural Economics, University College of Wales, Aberystwyth

LONGMAN
LONDON AND NEW YORK

Longman Group Limited
Longman House
Burnt Mill, Harlow, Essex, UK

*Published in the United States of America
by Longman Inc., New York*

© Longman Group Limited 1982

First published 1982

British Library Cataloguing in Publication Data

Haines, Michael
   Introduction to farming systems.
   1. Agricultural industries – Great Britain – Economic
  aspects
   I. Title
   338.1'0941 HD1925
   ISBN 0–582–45081–0

Library of Congress Cataloging in Publication Data

Haines, Michael R.
Introduction to farming systems.

Bibliography: p.
Includes index.
1. Agricultural systems. 2. Agricultural systems –
Great Britain. 3. Agricultural systems – Europe.
I. Title.
S513.2.H34 338.1'094 81–11833
                  AACR2

---

*Printed in Singapore by
Tien Mah Litho Printing Co (Pte) Ltd*

Contents

Preface

This book is based on an established and successful first-year undergraduate course originally devised for students in environmental science, but now taken by students in agricultural economics, agricultural marketing, business studies, economics, accounting, geography and sociology. It offers a basic understanding of agriculture for the growing number of people whose work brings them into contact with agriculture, but who do not need the detailed knowledge offered by texts written for specialist agriculture students. It is also aimed at a wider general public which is increasingly concerned about the economic and political weight of the agricultural lobby and the impact of the industry on the natural environment, but which has relatively little access to non-technical information of an unbiased character. The intention is thus not to provide detailed technical knowledge about agriculture and a dozen allied subjects (agricultural economics, soil science, etc.), but to develop understanding where there is frequently only misunderstanding and suspicion.

To this end, the book deliberately avoids technical detail, and is written in non-technical language. Complex issues are reduced to their bare bones (to the inevitable dissatisfaction of the specialist), but readers are encouraged to follow up this broad-based introduction through supplementary reading indicated at the end of the book.

Like all textbooks this one is inevitably derivative, and acknowledgement is freely made to sources too numerous to name. I would like to thank Professor D. I. Bateman, Department of Agricultural Economics, U. C. W. Aberystwyth, and Dr Andrew Speedy, East of Scotland College of Agriculture, for their very helpful comments on the draft text. Any errors which remain, however, are of course my own. I am also grateful to the following individuals and organizations for providing supporting material:

Brooms Barn Experimental Station: Figs 6.3, 6.4, 6.5, and Table 6.1
Crown: Figs 6.1, 6.6, 6.7
H. Davies: Table 7.1
Faber & Faber Ltd: Fig. 2.6
Farmers Weekly: Figs 2.2, 4.2, 6.2, 6.8, 6.9, 7.3, 8.3, 8.4, 8.9, 9.2, 10.5, 10.8, 11.1, 11.2, 11.4, 11.8

W. D. Jones and A. M. Sherwood: Table 8.1
Longman Group Ltd: Figs 10.1, 11.3
M.L.C.: Figs 8.8, 10.6, 12.2
Soil Association: permission to quote from 1976 pamphlet

All photographs, where not otherwise acknowledged, are by the author.

Michael Haines,
Aberystwyth,
March 1981

I

Introduction

Agriculture throughout the world is still man's single most important activity, for despite all the advances of high technology, it is still the only reliable source of food, and an important source of fibres and other products whose synthetic substitutes are often not as good as the natural products, or more costly to produce. In many countries agriculture is also the largest single employer. In many Third World countries it is the main or only source of livelihood for over 50 per cent of the population, and contributes roughly the same proportion to the national income. Its contribution to the economy of developed countries is also larger than is generally realized. In the USA agriculture employs more people than the car assembly, steel and transport industries put together. In Britain only 2.7 per cent of the civilian working population was employed in agriculture by the end of the 1970s, and it contributed 2.3 per cent of gross domestic product, producing 71 per cent of Britain's requirements for food that can be grown in this climate, which was an increase of twelve percentage points in a decade.

To the non-agricultural community, however, agriculture is an alien territory viewed from a car window. City dwellers in the past regarded it through rose-tinted spectacles as a Romantic occupation, but more recently they have come to regard it as a threat to the countryside. In Britain especially, modern agricultural methods have been increasingly criticized for their effects on the environment and their heavy dependence on non-renewable fossil fuels and on chemical fertilizers and pesticides, and from being an unknown territory agriculture has suddenly become everybody's business. The land which is agriculture's basic input and the farmer's workplace is the urban dweller's place of recreation, and the energy-dependent motor car has allowed townspeople to penetrate to the remotest corners of it. The result is that no other industry has been subject to such public scrutiny and criticism. At the same time, few industries have been so subject to interference and outright vandalism from some members of the public. Farms on the outskirts of towns regularly suffer damage to crops, fences and buildings from dedicated vandals or simply ordinary people walking dogs or taking short cuts, oblivious to the damage they do. Some lose crops and even animals to thieves intent on stocking the family freezer at the farmer's expense. In some areas livestock cannot be kept because gates are persistently left open or fences broken, since it is the farmer who is held responsible for straying animals even if a trespasser opened gates and let them stray. When foot and mouth disease ravaged Britain in 1967 (costing £27 million in compensation for 433,987 slaughtered animals) ramblers ignored statutory warnings to keep off agricultural land to avoid spreading the disease.

In this atmosphere of confrontation, in which each party sees the other as the villain of the piece, the real community of interest which unites the majority of farmers and the conservation lobby is obscured. There are unquestionably vandals on both sides, but it is no more reasonable to brand all farmers as despoilers of the countryside than it is to brand all members of the general public as thoughtless or deliberate vandals. The problem is that the non-farming majority of the population is generally ignorant of the way in which agriculture works, how it inevitably shapes the countryside, and how it is in turn affected by external pressures, not least of which is public opinion. This lack of understanding is a serious obstacle to an informed approach by planners and public authorities, as well as to mutual understanding between farmers and the rest of the community. The object of this book is to begin to remedy this situation by providing insights into agriculture for the growing number of specialists whose work brings them into contact with it, and for the public which has little access to objective, nontechnical information about the agricultural industry. The intention is not to champion modern agriculture nor to condemn it, but simply to explain practices which seem unnecessary or perverse when their purpose is not understood. Within the limits of one book it is clearly impossible to offer detailed information about agriculture and a dozen related subjects like agricultural economics, soil science, climatology, crop husbandry, etc., but that is not the point of the exercise. The object is to provide a meeting ground where there is currently too often only mutual misunderstanding, and readers seeking further information are referred to more specialist literature.

Modern agriculture is one of the most efficient and productive industries in the economy, producing more and more food on less land, with less labour. Farm hectarage in the UK declined by 8 per cent between 1900 and 1980, but cereal yields rose steadily over the decades. Trials in 1980 which compared the performance of some nineteenth-century varieties of barley with modern varieties growing in identical conditions showed that the modern strains yielded 42 per cent more grain. Earlier trials in 1978 showed that between 1947 and 1975 alone, the increase in yield was of the order of 31 per cent, and the latest evidence shows that in the 1970s all crop yields rose by 20 per cent. In an economy where industrial efficiency is judged in terms of labour productivity and optimum use of inputs, this was an impressive achievement, due very largely to improvements in plant and animal breeding, as these trials indicated. Agriculture's improved performance has also depended, however, on vastly improved technical efficiency, and on the use

Table 1.1 Increases in the production of some major agricultural products in the UK in the 1970s (Based on data in *Annual review of agriculture*, HMSO, 1981)

Commodity	Index of yield 1980 (1969/71 = 100)	Index of labour productivity (1975 = 100)	
	Per ha		
Wheat	134	1970	88
Barley	124	1972	100
Potatoes	126	1974	107
Sugar	98	1976	93
Oil seed rape	150	1978	116
	Per head	1980	129
Milk	125		
Eggs	112		

of modern aids like chemical fertilizers and pesticides, and agricultural machinery.

It is precisely the use of these modern aids which provokes criticism of modern agriculture, which is accused of becoming a branch of applied chemistry, with little understanding of the biological processes involved in plant and animal production. In reality, agriculture is a production process which has always exploited biological principles to produce food, fibres and other natural products, and income for the farming sector of the economy. The criticism is that agriculture today sacrifices the biological principles to economic pressures. Instead of working 'in harmony with nature', it is alleged, modern farmers consider the natural world as a resource to be exploited and transformed in the pursuit of higher profitability. The familiar stereotype of the ruthless agribusinessman bears little resemblance to most farmers, however, whose understanding of the biological principles of agriculture is after all generally greater than that of many of their critics. At the very least it is usually great enough to warn them how far they can manipulate the natural resources without seriously damaging the long-term viability of the farm on which their livelihood depends.

This book explores the interaction of the biological and economic factors which together shape farm systems. It examines why there are all-arable farms, all-dairy farms, or mixed enterprise systems, and how each is to some extent dependent on the others. The account is limited to the farming systems of temperate, developed countries (in Europe, North America, New Zealand, parts of Australia), and examples and statistics relate primarily to the UK. The term *farming system* is an old one, much older than systems analysis, which has recently invaded every field of study from engineering and biology to music and the arts. A system is simply an orderly set of interdependent and interacting components, none of which can be modified without causing a related change elsewhere in the system. The whole of the natural world is the biggest such system, composed of subsystems, each of which functions as though it were an independent whole although it is simultaneously part of other systems. Systems analysis has simply formalized techniques for investigating these interactions in any situation, including agriculture. But agriculture has always been studied in terms of systems. Each farm is a highly organized, integrated set of operations, which exists in a complex of natural, social and economic environments whose interactions shape the individual farm system, and the intention in this book is to explore this without invoking the formal procedures of systems analysis. For readers who wish to do so, there is a reference to some appropriate reading in the attached bibliography.

To clarify the idea of a system, it is conventional to discuss it in terms of *objectives, resources* (or inputs), *constraints* and *interactions*. All organized systems are designed to achieve certain objectives. There may of course be only one, though more often there are multiple objectives whose simultaneous achievement causes interactions which affect the way in which the total system operates. Profit maximization, for instance, may be the farmer's objective, but it may be modified by a desire not to ruin the landscape, a desire which involves him in practices which reduce his profit. Objectives are achieved by employing resources within a framework of constraints, which can include limited resources as well as legal prohibitions, regulation of certain activities and, as we suggested, public opinion. Objectives, resources and constraints interact to produce a viable farm system, and for the purposes of analysis it is convenient to consider such systems as though they were static arrangements (in which case, once the optimum set of linkages were determined, it could remain fixed for all time). In reality, however, everything is constantly changing. Even if things themselves appear not to change, they change in their relations with other things. A farm may thus remain in livestock rearing, with very little superficial change in its internal system, for generations, but the social and economic environment is always changing, affecting the profitability and even the viability of the system in the long term. The recognition of and concentration on this dynamic interplay of factors are what characterize formal systems analysis, and since interactions take place in time, the analysis cannot neglect the historical dimension. This means that the study is always relative to particular conditions, however, so any generalization can expect to be modified in the light of those conditions.

It should be obvious without detailed knowledge of agriculture that a farm is such a system. It has objectives: to produce income for the farm family and food for consumers, and these objectives interact. It has resources: land, capital, labour, etc., and it suffers from constraints like soil conditions, climate, planning regulations, etc. The interplay of these factors shapes the farming pattern of an individual holding and the wider environment in which it exists – the national economy and supra-national systems like the European Economic Community (EEC) and world commodity markets.

The main resource of agriculture, which is also its main constraint, is the natural environment. Agriculture is the sustained effort to improve the

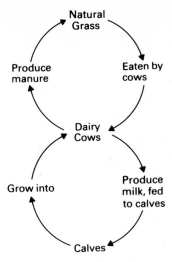

Fig. 1.1 A natural breeding cycle exploited by man

natural environment with the objective of providing food regularly and reliably. Livestock rearing is a systematic way of providing food without pursuing wild animals around the landscape. It exploits the natural breeding cycle of animals which is a completely self-contained, natural system (Fig. 1.1). This system is so simple as to be self-explanatory. A more complicated agricultural system which illustrates how man has intervened to improve on nature by eliminating some natural constraints is the Norfolk Four Course arable rotation. Devised in the eighteenth century, until very recently this remained the prototype of all rotational cropping systems around the world. It had been known for generations that the continuous growing of arable crops on the same land reduces soil fertility, thus reducing crop yields over the years. The only known remedy was to leave land fallow for one season in three, but this obviously reduced the overall level of output and thus limited the food supply available to support a growing population. The major advance which the Norfolk Four Course system represented was that it allowed land to be cropped continously without reducing its fertility. The succession of crops grown over a four-year rotation was winter wheat, turnips (fed to sheep), spring barley, and clover (fed to sheep). Every year a quarter of the land was sown to each of the four crops, the planting scheme rotating round the fields to ensure a systematic succession of crops to maintain soil fertility (Fig. 1.2). In two out of four years a crop was produced directly for human consumption – wheat for bread and barley for ale (the latter was long a staple of the diet). These crops impoverished the soil, and to restore fertility intervening crops of turnips and clover were grown for feeding to sheep on the field. This produced mutton for human consumption, and the animals' manure fertilized the field ready for the next cereal crop. Soil fertility was further increased by the capacity which clover has to *fix* nitrogen from the air (a process which converts atmospheric nitrogen

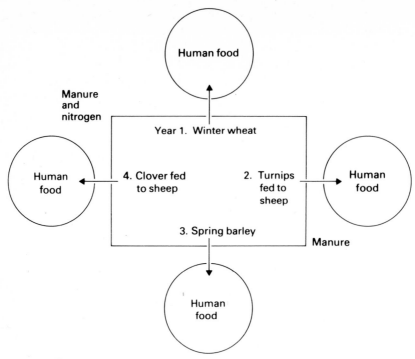

Fig. 1.2 The Norfolk Four Course rotation

into nitrate plant foods in the soil). Today the crops fitted into the rotation may vary, but the *principle* of rotation is still the basis of all mixed farming systems.

The history of agriculture is full of similar examples of ways in which biological constraints were overcome. There are many non-biological factors which constrain farming systems, however, including the availability of resources at the individual farm level. A lack of resources often restricts the range of enterprises to which the natural environment offers no constraint. It is clearly unwise, for example, for a farmer to venture into an enterprise with a high labour requirement if there is only himself and his wife to run it. Other constraints may be imposed by man-made conditions which make one enterprise incompatible with another. Sheep farming may not be possible on an intensive arable farm, which has large blocks of unfenced land, since animals have to be moved systematically around the farm as well as prevented from wandering.

Given the enormous range of options which result from the very different combinations of resources and constraints on any farm, not to mention the different motivations of farmers, the actual number of farm systems is very high. Within a given area there may be common constraints or resource problems which lead to a predominance of a particular farming system, but in the last analysis it is the individual's objectives and preferences which determine why one farmer keeps pigs while another does not. It is nevertheless useful to identify some broad types of farming system in relation to which individual farms can be better understood. The simplest classification is the division into livestock rearing and arable production which reflects man's early success in domesticating animals and improving strains of plants which he had previously gathered without cultivation. The earliest livestock-rearing system (still widely practised in the Third World) was nomadic, with animals being moved across vast expanses of natural grassland which man made no systematic effort to improve. The earliest arable systems were by contrast necessarily settled, since clearing the ground, sowing and harvesting are year-round activities, and even where people moved from year to year, clearing new patches for cultivation, it was invariably within a limited area.

From these two basic systems a whole range of more sophisticated farm systems developed, and their classification is still subject to dispute. It is possible to classify them by climatic region, in which case the major divisions are temperate, Mediterranean and tropical. This classification is useful in indicating some of the resources and some of the crops found in each region (e.g. sugar cane in the tropics, sugar beet in temperate regions). Whatever the geographical location, the stage of development of the local populations also affects farming systems, but like climate, this is a criterion external to agriculture, and a more useful one is the system of farming itself. Climate and stage of development will obviously be reflected in the farming system, but factors *specific* to agriculture in this case determine the classification. Dairying, for example, is identified as a system regardless of whether it is practised in a temperate or a tropical climate, by people in a developed or a developing economy, and only when a given system is described in detail must the impact of these external factors be considered. Although this book confines its descriptions mainly to the UK, the farming systems described are therefore fairly uniform throughout the temperate world, at least in relation to crop and animal husbandry, farm planning and broad management techniques, and allowances have to be made primarily for the timing of operations and the actual crops or animal breeds fitted locally into the general scheme.

The farm systems which this book identifies are: arable, dairying, upland grassland, mixed farming and intensive livestock. Though each is considered separately, they are by no means self-contained, and wherever possible their interrelations are brought out. The commonest link is that the output from one system becomes the input for another. The cereal output of arable farms, for instance, may become the feed input for dairy or intensive livestock enterprises.

One factor which links all farm systems is their economic management. All farming systems have the same twin objectives:

(a) to achieve an adequate level of return, measured in income or food for subsistence;

(b) to produce this regularly and reliably throughout a farmer's working lifetime.

This depends on maintaining over many years a consistent level of performance, using well-tried techniques and practices, but also showing constant responsiveness to new techniques which improve performance. In subsistence agriculture the primary objective is to maximize crop output, converting it into human and animal food with the minimum of waste. Once this level of subsistence is achieved, the farmer in developing as well as in developed countries becomes less interested in maximizing physical output than in maximizing his income, and this may tempt him to reduce output in the hope of raising the price. From the farmer's point of view this may be perfectly rational, since like any other workman he seeks as high a return as he can from his labour. It nevertheless conflicts with the consumer's equally rational desire to pay him as low a price as possible for his produce, so even in this most fundamental objective the farmer and the consumer may find themselves in opposition.

For the sake of simplicity, economists studying agriculture take profit maximization to be the farmer's primary goal, though in practice it is much qualified by other considerations. Consumers, after all, do not buy food solely on the basis of price, and profit is not the only motive which determines farm systems. Farmers' motivations and objectives are in fact no less varied than those of other population groups. Surveys suggest that in addition to the rational desire to maximize the return from their investment of capital and effort, most farmers are motivated by a strong desire to retain their independence, which they frequently value more highly than increased profits. Though few of them articulate the point except when pressed, nearly all of them also have an attachment to the land which they hesitate to call a *love* of the land, though the strength of the sentiment is revealed when that land is threatened. The economist's model of the farmer maximizing profits is thus an oversimplification, and in practice most farmers do not seek profit maximization so much as a satisfactory income.

To achieve this they need management skills as well as technical expertise, to ensure that they produce the right product in the right amount, with the right amount of inputs, and market it efficiently in order to generate the income they require. This exercise requires a good deal of planning, based on quite detailed information and a lot of informed guesses about the future, and as in any occupation group there are farmers who rise better to the job than others. Since there are often dozens of enterprises which could be combined in different proportions, each using different combinations of resources, the complexity of the planning function can become so great that it cannot be solved even with the help of the most sophisticated computers. Fortunately, however, on most farms the majority of enterprises are in fact excluded by factors like climate, soil type, and labour or capital availability. The choice is therefore more limited, which is in itself one of the major factors which prevents farmers from changing their farm system to seek higher profits.

Plan of the book

Though the objective of achieving a satisfactory income may be taken for granted whatever the farm system, it is clear that the resources available to achieve this objective, the constraints and system interactions will differ on every farm. There are nevertheless some general factors like climate, soil structure – in short, the natural environment – which determine many farming systems. It is these natural factors which set the most enduring limits to agricultural production, so it is with these that the book begins (Ch. 2), working towards the individual farmer who may, using ingenuity and capital in the form of technology, overcome some of the limitations of the natural environment. There is a cost involved in this, however, and whether farmers should be allowed, and even encouraged to push back the natural limits is a question which society has to answer, generally through the voice of governments. The socio-political and economic environment is therefore next to be considered (Ch. 3), followed by the resources and opportunities which shape farm systems (Ch. 4). It is at this stage that economics becomes important in determining which of the systems that are technically viable on a given farm is finally chosen, and how the farm business is subsequently managed (Ch. 5). On this foundation the remainder of the book then describes the principal farming systems found in temperate regions (Ch. 6–11). In each case a special topic is introduced to illustrate how a particular farm system may affect, and be affected by, the wider environment in which it exists. The presentation of these special topics is not intended to be exhaustive. Their purpose is to introduce a subject for further discussion and investigation in the class context, or for further reading. Chapter 12 discusses the marketing of farm produce, and Chapter 13 some likely future developments, including the increasing use of non-agricultural foods which may undermine agriculture's predominant position as our major food supplier.

What this plan has to neglect is the individual psychology of farmers, their motivations, inclinations and expectations, which are as important as any technical or economic factors in the shaping of farm systems. They show the same range of abilities and sense of responsibility as any other sample of human beings, and are no more and no less reasonable than the rest of the community. Like everyone else they seek a just return for their professional skills and their labour, but unlike most other occupational groups they remain essentially individual entrepreneurs, and the erosion of their independence concerns most of them more than the erosion of their income. This independence expresses itself primarily in the right to determine their own working conditions and, within the limits imposed by a highly uncertain industry, their own financial return and other non-financial rewards. Even when they are left free to make perfectly rational choices, however, often sanctioned and encouraged by governments, they invariably find themselves in conflict with the rational objectives, resource priorities and activities of the majority of the urban population. Though the farming population is small beside this urban majority, its interests are represented by organizations whose influence is undoubtedly disproportionate to the size of the community they represent, and it often seems to the rest of the community that the farming lobby has the ear of government, and is flatly opposed to the claims of other interest groups. One object of this book is to show that this apparent self-interest is frequently motivated by the quite reasonable conviction that the rest of the community simply does not appreciate the complexity of the farm business. The new image of the farmer as a self-centred, aggressive agribusinessman, indifferent to anything but his own profits, is as unrealistic as the old Romantic prototype with straw in his hair. The reality which should become clear is that efficient agriculture demands as much planning and management skill as the most complex industrial undertakings, and that most farmers are more concerned with preserving the land than they are given credit for – for the simple reason that their livelihood depends upon it.

2

The natural environment

The basic agricultural process is the production of plants by the conversion of chemical elements into carbohydrates, proteins, vitamins, fats, etc. The process is obviously dependent on the natural environment, and how far this has been modified by human activities.

The natural factors important to agriculture are climate, soil, air, and the genetic potential of plants and domesticated animals. Each of these is a resource exploited for agricultural purposes. The warm, moderately wet climate of north-west Europe, for instance, is ideal for grass production. The rainfall is a naturally occurring, cost-free resource which in arid regions would have to be provided by man, thus increasing production costs. The high rainfall therefore represents an opportunity capable of being exploited. Conversely, it would represent a constraint on the growing of cereals in these areas, since it might entail expensive land drainage and grain-drying facilities which would not be necessary in drier regions. The natural environment may thus be simultaneously a resource and a constraint, and the history of agriculture has often been associated with the attempt to reduce the impact of a natural constraint in order to seize the opportunity to exploit a natural resource. Without water, for example, the high temperatures of the semi-arid tropics cannot be exploited, but once irrigation is possible these regions of high solar energy can be used for food production.

CLIMATE

Climate affects agricultural systems on two levels. *The macro-climate* (i.e. the climate of a region) affects the choice between potential crops which fulfil a similar nutritional role, e.g. wheat or rice, since unless the crop is adapted to the existing climatic conditions it will not achieve economic importance in the farming system. The *micro-climate* is the climate within the crop itself, which affects not just the crop plant directly, but also the pests and diseases which attack it. The climatic factors which affect agriculture are temperature, moisture, sunlight and wind, all of which are interrelated.

Temperature

Every plant has a range of temperatures in which it grows satisfactorily, and these differ not only between crops, but also at different stages of the same

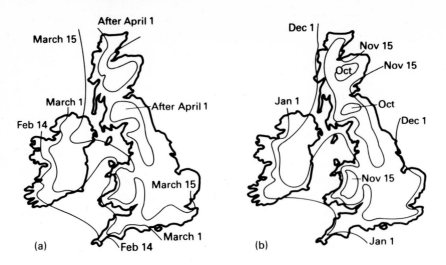

Fig. 2.1 The variation of the growing season in the UK
(a) The start of spring (daily mean temperature > 6°C)
(b) The onset of winter (daily mean temperature < 6 °C)

plant's growth. Many plants can survive freezing during their dormant period (e.g. fruit trees and cereals), whereas others are killed by frost at any stage of their life (e.g. potatoes and tomatoes). Others appear to benefit from a cold snap (e.g. celery, brussels sprouts).

Another factor determining the choice of crops is the length of the growing season, defined generally as the length of time that the soil temperature at a depth of 30 cm remains above 6 °C. This period depends on the latitude, altitude, slope and aspect of a particular piece of land. Even in a country as small as England the growing season ranges from 189 days at an altitude of 300 m in the northern Pennines to 322 days in lowland Cornwall (Fig. 2.1). A south-facing slope has a growing season 20 days longer than a flat site, and 40 days longer than a north-facing slope at the same location. The length of the season decreases with altitude, but the rate of decrease varies from 15–20 days per 100 m in northern England to 30–35 days per 100 m in the south.

Moisture

Soil water is essential for crop production, and this depends on the provision of an adequate water supply. An adequate supply is one which is above the plant's wilting point and below field capacity, which is the level at which a saturated soil loses water under gravity. Under natural conditions rainfall directly supplies all the water available to plants, and since the correction of any shortfall by irrigation became technically feasible in developed countries, rainfall there has not been a significant constraint on agricultural systems. Irrigation can be a very expensive procedure, however, so even where it has been technically feasible to irrigate land in other countries, the cost of doing so has inhibited agricultural development. In most temperate climates there is sufficient annual rainfall for the growing of most crops which are otherwise suitable, though in occasional drought years (like 1976) yield is reduced. In many years, however, there is a seasonal shortage of rain in the summer months which makes the cost of irrigation worthwhile in some areas (e.g. east and south-east England, north-east France and northern Germany).

The humidity of the air is not directly important to crop production, but a high relative humidity does influence the build-up and spread of fungal diseases of crop plants, especially mildews and potato blight. The spread of the latter is indeed predictable from information on relative humidity and temperature, and farmers are warned to carry out precautionary spraying against its spread on the basis of such predictions.

Solar radiation

Solar radiation is the energy source used by plants for the process of photo-synthesis and plant growth, and a shortage of radiation produces plants with a high stalk-to-foliage ratio, and with under-developed roots. In conditions where temperature, moisture and nutrients are in adequate supply, the theoretical efficiency of the conversion of visible radiation is about 5–6 per cent with only 1 per cent of the radiation appearing in the final product.

The amount of radiation falling on a piece of land is affected by cloud cover, which reflects back or absorbs some of the energy. Day length also affects the amount of radiation available to plants, and crop growth is closely related to day length. Since all points on the earth's surface receive the same number of daylight hours in a year, perennial plants derive no benefit from growing in one latitude rather than another, but this is not true of annual plants, which are generally grown in the period of longest daylight hours. Some crops achieve their optimum or crop maturity during periods of short days (e.g. barley, oats, wheat). Animal breeding cycles (and therefore animal production) are also affected by day length. The onset of oestrus in sheep, for instance, is induced by shortening day lengths.

Wind

The effect of wind on agriculture can be beneficial, as in the case of the prevailing south-west winds which bring warm ocean currents from the Gulf of Mexico across the Atlantic to Europe's west-facing coasts. These considerably lengthen the growing season in the coastal areas, allowing the production of early crops in Brittany, south-west England and Wales. Conversely, winds can have very damaging effects for agriculture. Wind destroys between 5 and 15 per cent of crops in the UK every year, and on some sites the losses can mount to 40 per cent. In addition high winds can cause soil erosion on light soils, prevent the activity of flying pollinating insects (thereby reducing yields), and cause the spread of weed seeds and diseases. Winds also carry damaging salt spray onto coastal crops, or industrial pollution which may prevent the growth of some easily-damaged crops or the grazing of livestock in the vicinity of industrial areas.

The damaging effects of wind can be reduced by planting shelter belts of trees, or very special planting techniques which exemplify the kind of precautions which have been developed over many generations to offset the worst effects of climate or climatic deficiencies. Thousands of years ago irrigation techniques were devised, sometimes on an immense scale, so that crops could be produced in areas suffering from an absolute shortage of rainfall or from seasonal shortages. In temperate regions today animals have to be housed in winter. Tropical plants have been introduced into temperate regions where they are cultivated in glasshouses where light and heat can be artificially regulated, and tender crops everywhere are protected from wind or pollution damage. In all these cases capital is used to overcome climatic problems, increasing the cost as well as the effort involved in the food production process.

SOIL

For most agricultural purposes soil is the indispensable growing medium for plants, providing physical support and nutrients. It has been replaced in some cases, however, by other inert growing media including peat and vermiculite clay granules to which nutrients are added in very controlled conditions. An extension of this is hydroponics, in which water alone is used to support plants. Although the technique is still only used for relatively high-value crops (mainly horticultural crops) it is almost certain to become more widely used as pressure on land increases, and it has proved capable of producing very high yields of high-quality food.

Soil is the product of the weathering of different types of rock and the incorporation within this material of organic matter. The quality of the resulting soil is affected by the climate in which the rocks were weathered (e.g. high rainfall leaches mineral nutrients from the soil), and the interaction of this climate with any vegetation which has grown on the parent material. It is indeed the climate more than the underlying parent material which generally determines the quality of the land.

Soil is a living medium, an ecosystem whose complex chemical and biological interrelations are still only imperfectly understood. As well as minerals it contains enormous quantities of living micro-organisms (fungi, bacteria, earthworms and insects), many of them beneficial and others harmful to plant growth. There is in addition a large amount of decaying and decayed organic material – dead micro-flora and fauna, plant roots, and material incorporated by the farmer (compost and farmyard manure). The micro-

Fig. 2.2 Lettuces grown commercially on a nutrient film without soil, producing very high-quality, uniform plants in tightly controlled conditions (Farmers Weekly)

organisms live on this decaying material, breaking it down into humus, and transporting nutrients to the plant roots. Humus has the capacity to hold plant nutrients which might otherwise be washed out of the soil, and it adds bulk to the mineral particles, thereby stabilizing soil structure. Its bulk keeps the pore structure open, allowing air and root penetration and efficient drainage, while simultaneously holding sufficient moisture to support plant growth. Soil fertility is closely related to the capacity to hold and make plant foods available, and it is because humus has this capacity that 'organic farmers' rely heavily on the humic content of soil to replace chemical fertilizers (Ch. 10).

Soils are grouped in zonal types sub-divided into soil series, in turn subdivided into soil types which indicate the kinds of crops that may successfully be grown on them. These soil types differ with respect to their depth, nutrient status, and their structure and texture.

Soil depth

Shallow soils occur widely in heavily glaciated areas and on slopes eroded by wind and water. They limit the depth of mechanical working and the rooting of crops, and they tend to be easily eroded. They have less bulk per hectare, and thus dry out more quickly than other soils. They are therefore most frequently sown to permanent grass and utilized for grazing. Deep soils, on the other hand, have all the opposite characteristics, and are therefore used for arable systems. The depth of usable soil can be increased by ploughing, and this is particularly important for some root crops, which require deep soil if high yields are wanted.

Nutrient status

With rare exceptions, soils in their natural state will not produce worthwhile crop yields without man's help. All plants require adequate supplies of chemical nutrients which they derive in solution from the soil. The amount of nutrients easily available to plants is affected by the relative acidity of soils, which may be naturally acidic or alkaline, depending on the underlying rock from which they derive and the degree of weathering that has occurred. Acidity reduces the availability of some chemical ions, important for plant nutrition, and increases others. The effect of alkalinity is not fully understood, but it reduces the uptake of phosphates, a major plant food, and affects the metabolism of iron. Soil acidity is measured in *pH*, a low number indicating high acidity, while 7.0 indicates neutrality.

As Fig. 2.3 shows, different plants tolerate varying degrees of acidity, but in practice it is generally sufficient for a farmer to maintain a soil slightly on the acid side of neutral, which suits most commonly grown farm crops. This is readily achieved by the application of naturally occurring limestone, an application which needs repeating roughly every five years in high rainfall areas. The application of lime also affects the micro-organisms

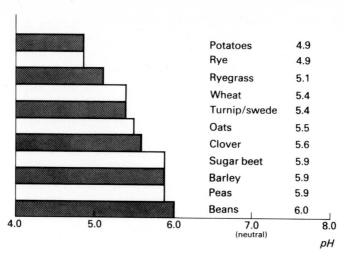

Potatoes	4.9
Rye	4.9
Ryegrass	5.1
Wheat	5.4
Turnip/swede	5.4
Oats	5.5
Clover	5.6
Sugar beet	5.9
Barley	5.9
Peas	5.9
Beans	6.0

Fig. 2.3 Relative tolerance of farm crops to soil acidity

in soil. For instance, it encourages the organism which causes potato scab, so liming before a potato crop is avoided. Conversely, it discourages the fungus which causes club-root of brassicas (cabbages, cauliflowers, swedes, etc), so it is commonly applied before these crops are planted.

Three major nutrients are necessary for active plant growth: nitrogen (in the form of nitrates), phosphorus (in the form of phosphates), and potassium (potash). Nitrogen accelerates plant growth by increasing the leaf area, which accelerates photosynthesis, and it is lack of this element in the natural state which limits crop growth. Nitrogen enters the soil from many sources. Plants absorb some from the air, but most enters the soil in rainwater as ammonia. Ammonia is also produced during the decomposition of organic material by soil micro-flora. However, all nitrogen compounds are exceedingly soluble, so they are quickly washed down the soil profile (see below) and lost for plant growth. One natural method of increasing the nitrogen content of soil is to grow legumes (peas, beans, clover, alfalfa). These are provided with sufficient nitrogen for their own growth by bacteria normally associated with their roots which have the ability to fix nitrogen. Ploughed into the soil they therefore increase its nitrogen content for the succeeding crop, as in the Norfolk Four Course rotation. Further nitrogen is added in the form of manure, which is initially rich in nitrogen, and in the form of manufactured nitrates. It is the use of manufactured nitrates which has greatly increased crop yields in the developed world, but their high cost has prevented their use in the developing world. The cost increased sharply following the oil price rises of the 1970s, since nitrogen fertilizer is manufactured from natural gas by a process which has a high energy input, and this increasing cost has reinforced existing doubts even in the developed world about the wisdom of over-reliance on manufactured sources of nitrogen in the long term.

Phosphates stimulate root development and crop ripening and improve crop quality, while potash is essential for photosynthesis. The presence of both in soils is directly related to their presence in the rocks from which the soils derive. Under natural conditions potash is not normally deficient, but phosphate deficiency affects large parts of the world. Both can be applied artificially to soils, and the application of phosphates can have particularly dramatic effects, reducing dependence on high applications of nitrogen to achieve high yields. Phosphate applied to clover, for example, produces a better crop, which in turn adds more nitrogen to the soil.

Besides nitrogen, potash and phosphates, there are many trace elements which plants (and the animals which graze them) need in minute quantities if they are to thrive. They include sulphur, magnesium, iron, manganese, copper, zinc and boron. In particular locations there may be patches of deficiency of one or more of these, but once identified the deficiency can be corrected fairly easily by appropriate applications. Certain plants also need large amounts of particular elements, and these also have to be added to the soil. An example is sodium for sugar beet. In excess amounts these trace elements may be harmful, however, and even lethal. There are pastures where animals suffer from excess molybdenum in the herbage, and sea-inundated land may have an excess of sodium which destroys the soil's permeability. These trace elements build up by natural means from the parent material, or the excess can be man-induced or man-accelerated by industrial effluents. Further problems are caused by interactions between trace elements which may be difficult to detect and treat.

Plant and animal products regularly remove mineral salts from the soil, some of which are returned as plant debris (dead leaves, roots, straw) or in the urine and faeces of grazing animals. In non-intensive farming systems these two processes can maintain a balance, but as production is intensified more is removed from the land than is returned, and the resulting deficiency must be made good if permanent depletion is to be avoided. One purpose of the traditional fallow period was to allow time for the soil micro-flora and fauna to degrade this organic matter, but this meant that land was periodically taken out of production. The introduction of inorganic fertilizers in the mid-nineteenth century allowed farmers to keep land permanently in production and simultaneously maintain soil fertility. There are experimental fields where continuous cropping over a century has been carefully monitored, and yields have not significantly declined.

Assuming a soil which is not deficient in any one element, fertility can be maintained by supplying nutrients in an amount equal to that removed in plant products, in drainage water, and made unavailable to plants by chemical reaction in the soil. Where some heavily fertilized crops are grown (e.g. potatoes) some nutrients which remain unused by that crop contribute to the growth of succeeding crops. The key to sound fertilizer policy on a farm is thus the determination of how much nitrogen, potash and phosphates are removed by each crop, and therefore how much needs to be applied to retain fertility. The farmer has access to published figures which provide this information for various yield levels of various crops, the fertilizer recommendations being expressed in terms of kilograms per hectare (kg/ha) of the nutrients, as in Table 2.1.

To supply this level of fertilizer a farmer can use compound fertilizer which contains all three main elements in different proportions. Alternatively, he could use *straight* nitrogen, phosphate or potash fertilizer, or a combination of straight and compound fertilizers. In addition he generally uses farmyard manure or compost where they are available, since both improve the soil structure and texture as well as adding nutrients. The choice will be influenced by operational principles too. For instance, little nitrogen is applied in the autumn when plant growth is reduced by low temperatures, since it will have been washed out of the soil by the time plant growth resumes in early spring. This is not true of phosphates and potash, which are retained in the soil and released gradually as plants require them. The normal practice is therefore to apply a low-nitrogen compound to autumn-sown crops, supplementing this with a straight nitrogen fertilizer in the spring.

The choice is also affected by the availability of manure and compost, and the price of purchased fertilizer. This cost varies considerably, depending upon the form in which the fertilizer is purchased. Basic slag is a cheap source of phosphates, but contains much less phosphate (16%) than concentrated *triple superphosphates* (50%), so much higher applications are needed to achieve the same effect. Though it is therefore relatively easy, given the chemical analysis of a fertilizer, to determine the cost per kilogram of nitrogen/phosphates/potash compared with other sources, the cost of application also has to be added, as well as other costs which are less easy to quan-

Table 2.1 Typical fertilizer recommendations for some common farm crops

Crop	Nitrogen (kg/ha)	Phosphate (kg/ha)	Potash (kg/ha)
Winter wheat			
at sowing	0–25	50–75	50–75
in spring	50–150	0	0
Spring barley	60–90	50–75	50–75
Potatoes			
without farmyard manure	150	200	250
with 40 tonnes/ha farmyard manure	100	150	150
Sugar beet (no manure)	125	63	125 + sodium
Winter oilseed rape			
at sowing	50	80–90	160–250
in February	150–200	0	0
Peas	0	0–50	0–150

tify. It is cheaper, for example, to use farm labour to spread a solid fertilizer than it is to employ a contractor with special machinery to apply a liquid fertilizer. At very busy periods of the year, this extra cost may nevertheless be worthwhile, because on-farm labour is then at a premium.

Soil texture and structure

Soil texture and structure, unlike the nutrient status of the soil, are both difficult and expensive to modify. By texture is meant the smoothness or grittiness of soil, which depends upon the relative proportions of the variously sized particles of which it is composed. The scale varies from stones and gravel (> 2 mm diameter) down to clay particles (<0.002 mm). In addition a very small percentage of soils classified as *organic* contain a high proportion of peat. This textural scale aids the farmer by indicating the soil's ability to retain moisture and to supply and retain nutrients, as well as indicating its heaviness or ease of working. It also suggests how stable the soil is likely to be in extreme conditions. For example, will it blow in high winds? What will be the effect of prolonged drought? Under heavy rain, will it turn to mud which dries as an impenetrable crust?

Soil structure and porosity are a function of the arrangement of the solid particles and the spaces between them. Structure is normally examined by studying a vertical slice through the soil, termed the *soil profile*. A well-structured soil is one which allows the ready penetration of plant roots and the free movement of air and water. This means that there must be no impenetrable layers which impede this movement, and one reason for soil cultivation is to ensure that such conditions are produced. The structure of soil is related to its texture, since if it is composed of very small particles these pack readily together to form impenetrable layers (called *pans*), and to its organic content.

Although it is possible to classify soils with very great precision, farmers use a more pragmatic classification which distinguishes seven types, each possessing characteristics useful in terms of their cropping potential.

1. **Sandy soils** are light, excessively free-draining, and therefore usually very poor, since any nutrients are quickly leached away. They may blow during periods of high wind, and very special precautions may have to be taken to prevent this (often special planting patterns). They are easily-worked soils requiring relatively little tractor power to cultivate them, but they are very abrasive, so the problem of wear on machinery can be serious. The worst sands are really only suitable for growing drought-resistant conifers, but on better sands rye, oats, barley, potatoes and carrots may be grown. Reasonable yields depend on the provision of adequate supplies of nutrients and water, irrigation being especially important for root crops. Given adequate water and nutrients, however, the ease of working and earliness of these soils make them particularly suitable for the production of market-garden crops.

The improvement of sandy soils was traditionally achieved by *marling*, which is the addition of clay, but this has become too expensive. Where possible, heavy dressings of manure or compost are instead applied which help the soil to retain water and nutrients for plant use. Green manuring is also widely practised in northern Europe, a crop of mustard (or other) being grown specially for ploughing in, in an effort to increase the humic content of the soil.

2. **Clay soils** represent the other end of the spectrum. They are heavy (or *strong*) soils, and the tractive effort necessary for their cultivation is 2 to 4 times that required for sandy soils. They are sticky, and any ploughing or cultivation has to be carefully timed to coincide with relatively dry periods, to avoid compaction (or *panning*) of the soil. Their ability to hold moisture is also an advantage, however, in that plants grown on them rarely suffer from drought or leaching of nutrients. They are nevertheless slow to warm up in spring, and where they are sown to grass for grazing, animals have to be removed in winter to prevent damage to pastures (called *poaching*).

Clay soils are normally rich in potash but deficient in phosphates, and they may need lime. Because of their ability to resist drought, and the high cost of drainage, a high proportion of these soils is devoted to grass production. Arable farming is possible where drainage has been carried out, though a period under grass is nearly always included in the crop rotation to improve soil structure (Ch. 6). The crops commonly grown on clays are winter wheat, oil seeds and beans.

3. **Loamy soils** contain varying proportions of clay, silt and sand, and range from fairly heavy clay loams to fine sandy loams. From the point of view of soil quality and management they present few problems. They warm up quickly in spring, are fairly drought-resistant, and retain nutrients added to them. They may require draining, but this is usually easy. There are very few restrictions on the crops they will support, so in this case the choice of enterprise is determined by factors other than soil characteristics.

4. **Silt soils** are characterized by a high proportion of very small silt particles which tend to pack loosely together, impeding the movement of air and water. Poor drainage is their main disadvantage, so they are best left under permanent grass. Confusingly, the alluvial 'silts' found around certain estuaries are really silt loams, and are very productive soils indeed. They can be cropped intensively with a wide range of horticultural crops, bulbs and field vegetables, in addition to cereals, sugar beet and potatoes. They are nevertheless rather unstable soils, and the inclusion of intermediate grass break crops and the addition of humus are desirable to maintain a satisfactory structure.

5. **Organic soils** at their best can be the most fertile soils found anywhere. At their worst, they occur as acid peat soils in upland areas which have limited agricultural value unless they are extensively and expensively improved.

The best organic soils occur mostly below sea level, as in East Anglia, so strong sea walls and pumped drainage are necessary which add considerably to the cost of using them. Moreover, since they are almost 100 per cent organic matter, they are readily oxidized away once drained and exposed to the air. In the Fens of eastern England the soil level is falling at a rate of 2 cm per year. Blowing is also a serious problem, and costs are increased by the need to add substantial quantities of potash, phosphates and trace elements, though these soils are rich in nitrogen. They can be used for almost any crop, wheat, potatoes and sugar beet being the commonest, but crops of peas, celery, carrots and other vegetables are also taken.

6. **Calcareous soils** are derived from chalk or limestone, and may contain varying proportions of sand or clay. If the clay content is considerable they tend to be difficult to work, but most are loamy in character. They may be very shallow, thus drying out rapidly, and they tend to be deficient in phosphates and potash. In the past these soils were devoted to grass for sheep farming, but the use of artificial fertilizers has allowed the growing of cereal crops, principally barley. Continuous barley cropping is indeed now common on these soils, only the poorest and steepest land being left as permanent pasture. This is true of large parts of England (e.g. the Cotswolds and Yorkshire Wolds), and the changed use of these areas as the result of chemical fertilizers and high cereal prices offers one of the most striking modifications in the agricultural pattern and landscape of the British Isles in the twentieth century.

7. **Soils of upland regions** may be any of the preceding types, but their productive potential is determined less by the soil type than by the climatic conditions of the uplands. High rainfall causes rapid leaching of mineral nutrients and drainage problems leading to the creation of areas of acid peat bog. These upland soils are thus generally devoted to grassland, which can be maintained with a minimal use of fertilizer. In some favoured areas barley can be grown, along with oats and seed potatoes, but a high proportion of the uplands is covered with plantations of trees because they will not even support grass without extensive improvement.

As in the case of climatic deficiencies, agriculture has a long history of improving the natural state of the soil which would otherwise restrict production. In high-rainfall temperate climates, the commonest major operation which involves farmers in considerable effort and expense is land drainage. Excessive water in the soil excludes air, which may cause plants to become stunted or die. The level to which the soil pores are saturated with water is called the *water table*, which fluctuates over the year as rainfall fluctuates, but should not ideally rise above about 0.5 m from the surface. On all but the most free-draining soils or steepest slopes this may therefore mean that field drainage is essential.

Good drainage ensures higher crop yields because plant root systems are healthier. The presence and impact of diseases may be reduced, and so too may competition from other moisture-loving plants which invade the crop. Where animal production is concerned, good drainage reduces animal in-

Fig. 2.4 Mechanical laying of a continuous perforated plastic pipe to improve land drainage

festation by parasites which thrive in wet pasture, and allows more intensive grazing without risk of excessive poaching of the soil and destruction of the herbage. In all cases the use of machinery is easier, and wear and tear on expensive machinery is often reduced.

Drainage may be simply achieved by well-sited and maintained ditches, or more expensively by underground pipes or mole drains (narrow tunnels channelled in the subsoil). The choice will depend on cost and the relative difficulty of the operation, as well as the nature of the soil and its topography. It will also depend on the financial return from the enterprise for which the land is to be used. In a low-output livestock-rearing system, for instance, all that may be economical is a system of ditches running along the spring line, whereas more expensive methods may be economically viable on land to be used for high-value crops.

The impact of drainage on the landscape can be both obvious and insidious; obvious in the visible agricultural improvement, and insidious in the harmful effects it may have on wildlife and even the archaeological wealth of an area. Extensive drainage has reduced the wetlands which are the specialized habitat of certain bird species and small mammals. The drainage water may also contain quantities of nitrogen, and when it is discharged into water courses and lakes it stimulates the growth of algae which reduce the oxygen content of the water, and this threatens other water life. This process of eutrophication has become very serious, though in fairness it should be noted that pollution from sewage works is more responsible than farming. The same is true of pollution from industrial effluents, which are generally more concentrated than the run-off from farmland which contains undegraded agrochemicals. It is the omnipresence of agricultural effluents that threatens the natural environment, however, since they reach into even the remotest corner of the land while industrial pollution is largely confined to urban areas. The answer to agricultural pollution is not to impose more regulations on an industry which is virtually impossible to police, given its dispersed character, but to reinforce and multiply means of improving farmers' awareness of the impact which farming practices may have on the environment. Together with an active policy of designating areas where agricultural improvement can only be undertaken with special permission, this has more chance of success than a policy of prohibition and confrontation.

AIR

Under normal agricultural conditions plants derive all the carbon dioxide they need for photosynthesis, and animals all their oxygen supply, from the air. When animals are housed during the winter or in intensive units, ventilation must be adequate to remove noxious gases, or the animals' performance will suffer and their lives may even be endangered. This is especially true of controlled environment houses, in which forced ventilation is essential. Ventilation is also important in the greenhouse production of horticultural crops, and the air can even be enriched with carbon dioxide to accelerate photosynthesis, which accelerates the growth of a high-value leafy crop like lettuce.

In most circumstances, however, the air becomes a farmer's preoccupation only when it is polluted. Industrial pollution of the atmosphere may be in the form of gases which poison animals or plants (e.g. sulphur and nitrogen dioxide, carbon monoxide), or ash and dust particles which are inhaled and eaten by grazing animals and cause scorching of arable and vegetable crops. In some places the pollution may be such that certain enterprises have to be abandoned. Animals grazing downwind of some steelworks, for instance, can suffer from fluorosis, a debilitating condition which leads to early death.

On the other hand, some agricultural operations also cause air pollution which can be offensive to nearby residents, and they too may have to be abandoned in close proximity to residential areas. Examples are liquid manure spraying and the burning of straw stubble after harvest. Both are good farming practices, but even lyrical supporters of organic farming have been known to object to the aroma of farmyard manure when it drifts into suburban gardens.

GENETIC POTENTIAL

Agricultural production depends on the use of plants and animals genetically suited to their purpose. Put very simply, agricultural production provides a store of starch and protein in the form of plants and animals selected and improved by generations of experience and two centuries of scientific research. All the cereal crops which provide the staple diet of most of the world's population belong to one botanical family (the *gramineae*) which also provides all the grasses used in pastures for grazing animals. They were selected very early in man's history, because they produce large seed grains which are rich in starch and protein, and over many generations further selection has produced varieties which yield more grain, or resist diseases, or survive adverse weather conditions better than other strains. Many of the livestock used for human food also belong to one class of animals – the ruminants, which have a digestive system with four stomachs which can convert grass and other fibrous material indigestible by humans into digestible meat. The only important exception is the pig, which is monogastric, ie. has only one stomach. Many species of birds have also been selectively bred over the centuries to produce protein in the form of meat or eggs.

Given what already seems a rich store of species, it may seem that there is no need for a systematic programme of research and development dedicated to the improvement of plants and animals used in agricultural production. There are in existence plants and animals which are suitable for most of the ecological situations in which agriculture is practicable. In some situations the productivity of the species may be limited by climatic or other constraints, but certain species exist which can and do provide food for the small populations which live there. Even so, throughout man's history

Fig. 2.5 Experimental plots of barley varieties grown at a plant breeding station

efforts have never ceased to improve the productivity of plants and animals used in agriculture, and today more effort than ever is put into the breeding of new varieties and strains of crops and animals. The object of much of this work is to produce types which will thrive in previously unfarmed areas – for example, hardy wheats which will survive in the tundra, drought-resistant crops for the tropics, dairy cattle with improved heat tolerance for tropical climates. The other principal aim is to introduce resistance to pests and diseases, and a third objective is to increase the yield of both crops and animals to support the world's expanding population.

In all the research on plants, the introduction of genetic material from wild species by cross-breeding is important, supplemented by the highly skilled techniques of genetic manipulation. In animal breeding, cross-breeding from known highly productive strains has produced very significant increases in the yield of dairy cows, in the production of meat from pigs, of meat and eggs from poultry. Plant and animal breeders are often criticized, however, for showing more interest in increasing output than in increasing the efficiency with which production inputs are utilized. In a situation of surplus production such as exists in the developed countries of the west, increased output is not necessarily desirable, especially if it is achieved through the increased use of scarce resources, and it is unquestionable that many of the improved plants and animals produce more marketable output only if they receive more inputs than older strains. In the developing world where there is generally a shortage of inputs, these new strains are therefore unlikely to make a significant contribution to the production of more food. For this reason, emphasis is increasingly placed on the development of plants and animals capable of using less, and lower-quality inputs to better effect – for instance, animals which use low-grade feed instead of grain, or crops which produce reasonable yields without large amounts of artificial fertilizer. This is a question which is only now beginning to be debated in developed countries, where most of the research has traditionally been done, but it is certain that the objectives of plant and animal breeders will come increasingly under the scrutiny of the non-agricultural community and the governments which fund their research.

FARMING SYSTEMS OF THE UK

The way in which natural factors shape agriculture is illustrated by the distribution of farming systems in the UK. The dominant factor is climate, but climate is obviously related to topography. Figure 2.6 shows the distribution of the main farming systems of the UK in relation to a line drawn across the country from the Exe to the Tyne. To the north and west of this line is the highest land, which has higher rainfall and relatively warm winters. To the south and east is lower land, drier, with more extreme continental winters. This division is reflected in the predominant farm systems found either side of the line: livestock rearing to the north and west, arable and mixed farming to the south and east.

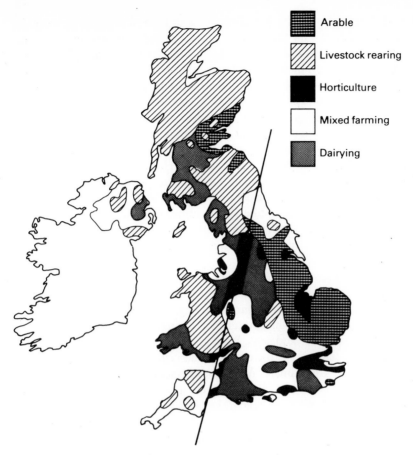

Fig. 2.6 Distribution of farming systems in the UK
(Adapted by permission of Faber & Faber Ltd from *Agricultural resources: an introduction to the farming industry of the UK*, Angela Edwards and Alan Rogers, 1974.)

This generalization says nothing about the detailed combinations of enterprises found on individual farms, nor about specialist areas which for various reasons – climate, soil type, aspect, closeness to markets, even history – have always produced specialist crops: flowers in the Scilly Isles, tomatoes in the Channel Islands; glasshouse crops near London; flower bulbs in Lincolnshire, imported centuries ago by Dutch immigrant farmers. It also neglects the interactions between different enterprises and different systems, of which the interdependence of livestock rearing and the production of animal feeding stuffs is a good example.

The arable areas of the country produce grain and some hay or dried grass which are shipped into livestock-rearing regions to feed stock over winter, when fresh fodder is unavailable. This makes settled agriculture possible in hill and mountain regions which in other countries cannot support livestock over the winter, the only solution being to move animals to lower pastures. The livestock-rearing areas in turn produce store stock for sale to lowland farms for fattening or breeding, and thus serve as a supplier of inputs for the meat industry throughout the country. This interdependence means that the uplands remain populated and farmed, while lowland farms are provided with the stock they need to integrate into a mixed arable and livestock system. The output of one system is thus the input to another, and changes affecting one system have repercussions for the other. If lowland farmers abandoned beef and sheep production for dairying, for instance, because milk brought a better return, they would not buy the store stock sold from upland farms for fattening. Hill farmers would therefore have to cut back stock numbers and lose income, or fatten the stock themselves. To do this they would have to plough up hills to grow more food on the farm, since they would not be able to afford more purchased feed. They might also have to provide sheds for winter housing and spring lambing, resulting in more expenditure and also very visible effects on the upland landscape.

The only farm systems which escape the immediate effects of the natural environment are the non land-using industrial systems which produce food intensively in controlled environments. These include some horticultural systems, but more importantly, intensive rearing of pigs and poultry. In the USA and Australia beef and some lamb are also produced in intensive 'feedlots' which it is claimed are the only commercially successful way of raising animals in a semi-arid climate. All of these systems nevertheless depend on agriculture for some, if not all their feed requirements, so until industrial processes can supply all these needs, they will remain locked into traditional land-using agricultural systems, and subject indirectly to the vagaries of the natural environment on which they depend.

3

The social, political and economic environment

Except in subsistence agriculture, farming is essentially an economic activity. It is one component in the economy of a nation, subject to changes elsewhere in the national economy and in the wider economy of the international community. It is sensitive to changes in the costs of inputs and the prices at which it can sell its output. It suffers from the effects of inflation, high interest on borrowed capital, high labour charges, and is frequently subject to political pressure to adapt to the economic needs of the nation. When there is a balance of payments deficit, farmers are urged to increase domestic production to save imports (in countries like Britain which are not self-sufficient in food) or to boost exports (in countries like France where production far exceeds home demand). In periods of industrial expansion agriculture may be regarded as a source of labour, so farmers are exhorted to increase productivity in order to release manpower for other industries. Even economic policies not directly related to agriculture can have far-reaching effects on the industry's structure. In Britain high rates of tax on unearned income, including farm rents, have contributed to an increase in owner-occupation and a decline in farm tenancy. Taxes on inheritance may compel the sale of part of a large farm to pay the tax due, and this may conflict with an agricultural policy whose object is to increase the size of farms.

Figure 3.1 indicates the complexity of the farming industry's relations with the socio-political and economic environment, showing how most factors affect agriculture indirectly as well as directly through their effect on a nation's agricultural policy. In this way it is not just the structure of the industry which is shaped, but even the choice of enterprises and management practices at the individual farm level. Since the individual farmer is rarely able to do much to change agricultural policy, in the short term it is he who has to adapt his farm system. In the long term, however, he may succeed in changing policy through his political organizations, which in all countries have a remarkable record of persuading governments that the public interest, not just their own, is served by supporting agriculture.

Like most government policy, agricultural policy is the result of a series of compromises between sectional interests – domestic producers, foreign trading partners, consumers, etc. – and it is profoundly influenced by social as well as political and economic considerations. Farmers constitute one of

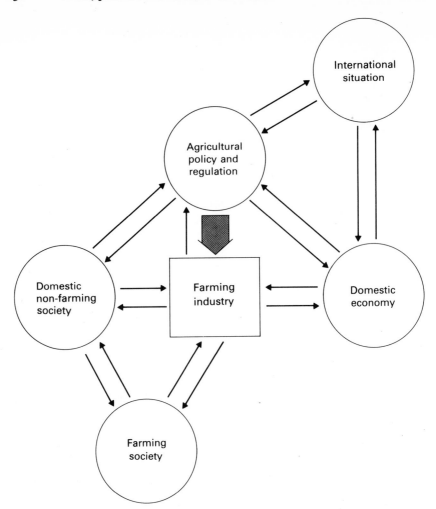

Fig. 3.1 Agriculture's relations with the wider community.

the most cohesive occupation groups in any society, whose personal values
have generally survived the impact of changes which have dislocated those
of other sectors of the population. Farmers' attitudes towards their own industry, their group loyalties and standards of behaviour have a profound
impact on their individual farm systems, and on the way in which their
organizations present their case to governments and the public. One factor
widely believed to mould their attitudes is tradition, reflected in stubborn
adherence to accepted farming practices and resistance to innovation. In
fact, few industries have shown such responsiveness to new techniques, and
where farmers do seem addicted to traditional practices it is generally not
out of stubbornness. Most of them willingly adopt new machinery and cultivation techniques which reduce their workload and increase productivity,
as the critics of modern farming would be quick to point out. They may

nevertheless by very reluctant to adopt totally new enterprises or change radically from one system to another, and where this is based on thorough knowledge of local conditions and the limitations they impose, or the farmer's own limited experience, it is prudent rather than reactionary. However, resistance to change among farmers may be as related to human psychology as it is in the rest of society. Fear of criticism by their peers and fear of failure can be as powerful a motivation as economic judgement. So too can normal human pride, pride for example in high crop yields achieved at almost any cost, or in livestock bred less for their commercial qualities than for the breeder's personal satisfaction.

One such factor which is particularly strong in areas where small farms predominate is fear of getting into debt. Again this may seem prudent, but in Europe in the past this fear inhibited modernization by preventing the adoption of improved production techniques and marketing methods, and in developing economies it is still a major obstacle to agricultural advance. Farmers everywhere are also highly independent characters who pride themselves on being their own masters. This expresses itself in a strong desire to own the land they work, and a consequent preference for owner-occupation even though tenancy may be financially more remunerative. Whether they own the land or rent it, however, farmers invariably have an equally strong desire that it should continue to be worked by their descendants. The motivation is simple: building up a farm can be a lifetime's work, and like most fathers farmers tend to believe that their work will survive better if it stays in the family. Both attitudes can obstruct restructuring programmes designed to increase farm size, which obviously means reducing the number of independent farmers. They also help to explain farmers' resentment of non-farming interests which intrude upon their independence. A classic example is the resentment of farmers whose land is designated part of a protected area or national park, even though it remains their property, since their freedom of action is restricted while the public is simultaneously allowed access to land which it regards as part of the national heritage.

The attitudes of the non-agricultural community have more effect on the farming industry than on many others, and increasingly influence the formulation of agricultural policy. These attitudes are basically ambivalent because the public has little firsthand knowledge of farmers, and knows them only through conflicting popular stereotypes. In many countries the farmer is still regarded as the 'salt of the earth', respected as the last repository of old moral values and self-sufficiency. The small farmer especially is believed to be hard-working, and indispensable to a thriving rural society. Conversely, he may be seen as a charge on public expenditure, since in most countries his income has to be subsidized by the rest of the community. In some countries he is now regarded as a despoiler of the countryside and cruel exploiter of animals, producing unwanted food surpluses at public expense, protected from the economic realities which shape other industries.

Despite recent criticism of this kind, farmers in most countries – with the notable exception of the UK – have retained an extraordinary degree of public support, for the simple reason that they produce the most basic commodity on which man depends for his survival. This has allowed governments throughout the developed world to pursue a policy towards agriculture which has been remarkably consistent, despite the wide range of national circumstances in which it has been formulated. The question might well be asked why governments have felt the need for a long-term policy towards agriculture when no such policy has seemed necessary for, say, the chemical industry. The answer is complicated, and since whole books are devoted to the subject, this chapter can obviously do no more than indicate the complex issues involved. The reasons are social and political as well as economic, and they hinge on three major considerations:

1. The relatively low incomes of the farm sector *vis-à-vis* other industries and social groups.
2. The strategic argument for maintaining a substantial domestic agriculture.
3. The economic contribution of agriculture to the balance of trade.

Governments acting in the name of the public have an obvious interest in ensuring that adequate food supplies are maintained at reasonable prices, and the failure to stabilize farm production and food supplies has fuelled political instability in more than one country in recent memory. The maintenance of food supplies and the support of agriculture have thus enjoyed higher priority than the support of any other industry except energy. Governments have sought to maintain food supplies by encouraging domestic production, or through trade policies which ensure that food is imported regularly and reliably from abroad. In Britain and Europe the case for self-sufficiency was strongly reinforced by food shortages resulting from two world wars which focused attention on the dangers of relying heavily on the world market. This strategic argument has been especially influential in the UK, which has imported roughly half its food supply throughout this century, and was twice reduced to strict rationing by wartime disruption of normal merchant shipping deliveries of food from overseas. Although British governments have always been active in negotiating trade agreements involving food imports, the main thrust of agricultural policy has therefore been towards the expansion, or at least the maintenance of a vigorous domestic farming industry, and a main plank of EEC farm policy is the same drive for self-sufficiency in Europe as a whole.

Another disadvantage of relying on the world market is that food purchases from abroad have to be paid for out of export earnings from manufacturing industry, and if exports are not high enough, food imports may have to be restricted. The threat was a real one in Europe after the Second World War, when industrial capacity was either in ruins or geared up to war production rather than the manufacture of exportable goods. In large parts of the Third World the threat is still serious, as the Brandt Commission warned when it argued for a more equitable balance of trade between the developed and the developing worlds. In the circumstances it is not surpris-

Table 3.1. The self-sufficiency of the EEC and some member countries in major agricultural commodities 1968/69 to 1977/78 (Based on data in *The agricultural situation in the Community*, EEC Commission, 1980 Report)

Commodity	EEC 9		United Kingdom		France		Germany		Italy	
	1968/69 (%)	1977/78 (%)	1968/69 (%)	1977/78 (%)	1968/69 (%)	1977/78 (%)	1968/69 (%)	1977/78 (%)	1968/69 (%)	1977/78 (%)
Wheat	94	105	45	65	154	185	86	98	95	78
Barley	103	106	99	107	158	163	80	84	22	35
Maize	45	52	0	0	149	119	12	18	47	61
Rice		67	0	0	45	10	0	0	225	172
Potatoes	100	99	95	92	102	99	96	92	97	99
Sugar	82	117	34	37	117	183	89	123	93	91
Fresh veg.	98	93	78	73	95	92	53	33	112	117
Wine	97	99	0	0	92	97	56	60	110	127
Fresh milk products	100	100	100	100	99	101	100	100	100	98
Butter	91	111	10	38	119	112	104	133	67	70
Eggs	99	100	99	100	100	98	87	76	94	96
Beef	89	96	61	73	111	107	87	101	65	60
Lamb	56	65	42	57	82	74	85	56	74	59
Pig meat	100	100	58	63	87	84	95	88	88	76
Poultry meat	101	105	99	102	102	115	49	59	99	98

ing that other governments have shared Britain's preference for the expansion of domestic agriculture in the search for self-sufficiency, or even to achieve an exportable surplus. To this end, price support and output control policies have been specifically designed to increase total production, and criticism of the farming industry for constantly increasing output at the expense of quality and the natural environment might be better directed at the governments which have consistently encouraged it, and occasionally obliged it to do so. These policies became common in Britain only after 1947, following a century of agricultural free trade. In most European countries such policies have been in operation for the whole of this century, however, and this divergence is one factor which has led to disputes over EEC farm policy, as a later section will elaborate.

The argument for expanding domestic agriculture has often been strengthened by the effects it may have on the rest of the economy. For instance, consumers in the west eat very little farm produce in the state in which it left the farm, but only after it has been processed. A large and efficient agriculture can therefore support a substantial processing industry which generates employment and contributes to the gross national product. Or again, if support for agriculture increases its efficiency, it should use fewer resources per unit of output, and thus release these resources for use in other industries. In many countries another economic argument for supporting agriculture is the desire to prevent rural depopulation, leading to urban overcrowding and the disintegration of the rural infrastructure of roads, schools, shops and other services. Agriculture is the largest rural industry, and governments seeking to prevent a general drift of population to the cities have supported farm incomes as an indirect means of sustaining the rural infrastructure on which the rest of the rural community depends. In some countries (e.g. France and Germany) this desire is so strong that a large number of small family farms have been heavily subsidized even though a smaller number of large farms might achieve higher productivity, social stability in the rural community being judged more important than strictly economic considerations.

In all countries, it is fair to say, the main reason why governments have supported agriculture is concern for the well-being of the farming community. This concern has often been motivated by political interest, since the farming vote in some countries has held the balance in national elections. Even in countries where the farming vote is less important, however, as it is in Britain, it has generally been accepted that without government intervention farm incomes would be low relative to the rest of the economy. A more accurate statement is that *some farmers' incomes* would be unacceptably low because of the size of their farm or its unfavourable location, or because their main enterprise is subject to particularly wide price fluctuations which result in substantial losses in some years.

One reason for this is the finite size of the human stomach. In the developed west where most people have enough to eat, as incomes have risen

Table 3.2. Household consumption of various food commodities in the UK 1955–79
(Based on data in *Household food consumption and expenditure*, Annual report of the National Food Survey Committee, HMSO, various years.)

Commodity	Household consumption per person per week (g)			
	1955	1960	1970	1979
Liquid milk* (litres)	2.3	2.3	2.1	2.4
Butter	126.7	161.0	169.5	126.2
Cheese	80.2	86.2	101.5	108.6
Margarine	132.7	103.8	81.1	102.9
Eggs (no.)	4.2	4.6	4.4	3.8
Sugar	500.1	503.5	480.2	327.4
Beef and veal		247.8	220.3	233.6
Mutton and lamb	516.8	187.9	146.8	120.8
Pork		57.3	79.7	102.4
Poultry	N.A.	47.6	134.7	184.3
Potatoes	1734.1	1621.6	1367.6	1157.5
Fresh green vegetables	419.3	448.2	337.9	235.3
Bread	1562.9	1289.0	1079.3	888.7
Frozen peas	N.A.	N.A.	28.9	49.6
Cakes and pastry	N.A.	178.9	126.4	80.8
Breakfast cereals	N.A.	51.0	77.7	95.8

* Does not include welfare milk.

there has been no comparable increase in the amount people eat (Table 3.2). They may have changed their consumption pattern from starchy foods to meat or other foods with higher protein content, but the total demand for food products has grown less rapidly than income. A gap has therefore opened between farmers' incomes and those of other sectors, and this would widen if governments did not intervene. This would happen even if the costs of inputs to agriculture and industry remained static, but since they tend to rise, the gap widens. Costs rise because the industries whose revenue is growing compete for resources like capital and labour which are used by farmers and firms producing farm inputs like machinery and agrochemicals. Efficient farmers respond by increasing their productivity, or by changing from low-revenue enterprises to others which bring a higher return. When they do so, however, they add to production levels, and the increased supply causes prices to fall. Since this fall in price does not bring about a corresponding increase in consumption, there is further downward pressure on their revenue. As for the numerous farmers who cannot adjust their farm systems significantly, they are left even further behind in terms of income. If they left the industry the incomes of those still in farming might keep level with other sectors, but many social reasons dissuade them from leaving. Older farmers would find it difficult to find alternative employment, and the cost of moving and strain of changing from a lifetime of self-employment can be very serious. Younger farmers might suffer less from the need to re-adjust, but the difficulty of finding alternative employment is still formidable.

Though these arguments have been widely accepted by governments, they are by no means unchallenged. For although farm incomes may be low in relative terms (and it is true that even with government intervention some sectors still suffer from low relative incomes), in absolute terms they may be quite high. This is certainly the case where changes in land values are taken into account, especially when they rise during a period of general inflation. The low relative income of an owner-occupier thus has to be offset against the appreciating value of his land, which is paradoxically increased further by government intervention to protect farm incomes.

The case for a price and output policy designed to guarantee farmers a stable and reasonable income has always been inseparable from the need to even out the worst effects of supply and demand fluctuations on consumers. The supply of many agricultural products fluctuates in the short term as the result of factors beyond the control of producers or governments (e.g. adverse weather, disease incidence, etc.). Where there is undersupply the only recourse may be to import supplies in order to hold down consumer prices. Where oversupply threatens to depress prices seriously, action is often taken to remove the surplus from the market in order to maintain the producer price at a reasonable level. In both cases it is again widely accepted that the farmer needs to be compensated for loss of income by the government. In some countries, however, it is normal for farmers to take out private insurance against losses arising from climatic extremes or similar calamities, and there is no fundamental reason why they should not do so elsewhere.

These short-term fluctuations may be generally unavoidable, but there are also production cycles extending over several years, with alternating periods of surplus and deficit, which result from farmers' expectations of future prices. The most notorious of these is the pig cycle. If pig prices are high and producers move into pig production in large numbers, or existing herds are expanded, the extra production reduces the price when the pigs reach the market. Pig production is therefore cut back, resulting in time in a shortage which increases prices, and the cycle is renewed. If producers had better information about other farmers' production plans, these cycles should in theory be avoided. However, in a fragmented industry composed of individual entrepreneurs in competition with each other, better information is unlikely to solve the problem alone. Governments have therefore adopted various policies to try to regulate these cycles as part of an overall policy of supply management.

GOVERNMENT POLICIES TOWARDS AGRICULTURE

Governments in most developed countries have devised policies to provide

farmers with a reasonable income and consumers with reliable supplies at reasonable prices, through a combination of the following methods:
1. Control of supplies reaching the market.
2. Manipulation of the price paid *to* producers or *by* consumers.
3. Structural reform of the agricultural industry.
4. Income supplements for farming families.
In addition governments have used less obvious policy instruments, like the hygiene regulations enforced in most countries to protect public health. These have frequently been used to control supply by excluding imports on the grounds that they were not produced, packed or processed under local public health regulations.

The most serious criticism of these policies is that they prevent the price from serving as a market indicator by communicating consumer needs to the farming industry. Lamb producers, for example, are paid not on the *quality* of carcases, but on their *weight*, so producers are encouraged to put extra weight on their lambs before sending them for slaughter. This extra weight is added in the form of fat, however, which consumers do not want. The pricing policy thus encourages farmers to produce a product which consumer research proves is unsatisfactory, and efforts to persuade farmers to market leaner lambs fail because the price they receive is not related to the proportion of lean, saleable meat. One of the main reasons for separating farm income support from pricing policy is therefore that the price would then become a truer reflection of consumer requirements, so farm output would have to be more closely matched to its market.

Control of supplies

In the developed countries the demand for most agricultural products shows very little growth over time. This is partly a result of low population growth, and partly related to the fact that consumption of food does not rise in line with incomes. Even price changes do not bring about very substantial changes in overall consumption. On the other hand, technical advances constantly increase agricultural output. The result is that when there is an excess of supply over demand, prices fall, and with them farm incomes. The converse is also true: when supply falls, prices rise, and so do farm incomes, since again there is little change in consumption. The object of government intervention is thus to regulate supplies in order to achieve the level of income which is judged desirable.

Supply control is easiest in countries which import food commodities, for imports are restricted and domestic prices rise to the predetermined level. This can be achieved simply by imposing a quantitative quota, or more usually by a tariff, or a variable import levy. In this last case the desired price is determined, and on this basis a minimum import price (MIP) is calculated which is frequently higher than the domestic price in order to give some preference to home producers. If foreign suppliers are prepared to supply a commodity below this price, the importing country adds a levy to raise it to the MIP. If exporters lower their price further, the levy is increased to maintain the import price on a par with the domestic price – hence the term *variable import levy*. One advantage of this policy to the importing country is that the levy revenue accrues to its exchequer, and it becomes particularly attractive if commodities can be bought in the cheapest world market, since this increases the levy revenue. The system nevertheless increases consumer prices, and it disrupts world competition as well as inviting retaliation from other countries. Moreover, once all imports have been excluded, it will not prevent domestic prices from falling further, so some additional action is needed, generally some form of control of the domestic supply. Measures of this kind are also needed in countries like the USA which are self-sufficient in food, or produce a surplus for export.

In these circumstances a government agency purchases enough of the commodity to raise the domestic producer price to the predetermined level. Procedures vary very widely, but the effect is always identical: supply falls on the market, consumer prices rise, and the agency acquires a stock which has to be disposed of later without reducing the producer price below the desired level. Sometimes this can be achieved within the home market, if on average supply and demand are in balance, and the surplus which is depressing prices is only temporary (as where there is seasonal surplus following a period of deficit). However, the procedure is often used to support the prices of commodities in long-term surplus, in which case the agency tries to sell its stock on the world market. If this market is in deficit there is no difficulty, but if the market is fully supplied at the prevailing price, it may have to dispose of stocks at very low prices (as in the case of EEC surplus butter stocks). The resulting loss is subsidized out of other funds, a procedure which may upset the world market and bring complaints of dumping from other exporting countries.

These methods are essentially short term, and they are used where a surplus already exists. One means of preventing a surplus from arising is a long-term policy limiting the resources available for the production of a commodity. This may entail the imposition of production quotas, a device used in many countries including Canada and Britain. Another method used in the USA to prevent grain surpluses entailed the payment to farmers of a sum in lieu of every acre they withdrew from production. In the EEC dairy farmers have been paid in a similar manner to transfer cows from milk to beef production. This method of controlling supply is effective, but costs almost as much as storing surplus production, since the payment to the farmer must equal the profit he could have made by continuing in production, or the difference between the profitability of the enterprise he abandoned and that which he adopted in its place. There have also been side-effects of such schemes which reduce their effectiveness. In the USA the production per acre from the land remaining in cereal production actually increased because more effort and fertilizer were allocated to the reduced area. Moreover, since only the worst land was taken out of production, large acreages had to be withdrawn in order to achieve worthwhile reductions.

Manipulation of producer and consumer prices

All forms of supply control pay the farmer at the expense of the consumer. They thus satisfy one policy objective – the protection of farm incomes – without ensuring reliable supplies at reasonable prices, which has been the other principal objective of agricultural policy. In most countries this reality has been accepted as inevitable, so policies have generally protected farmers without seeking to protect consumers. It is nevertheless possible to satisfy both objectives simultaneously where a country imports the supported commodity. Assuming that the world price for a given commodity is lower than the price desired for domestic producers, consumers may be allowed to pay the world price for imported and domestic supplies alike. Domestic producers are then paid a *deficiency payment*, which is an amount equal to the difference between the world price and the higher price judged necessary to maintain their incomes. The deficiency payments are met out of income taxes, and this has the advantage that low food prices are subsidized by high income earners, while low income groups are protected from the high food prices which everyone must pay under a variable import levy system. Both objectives are thus met, and in addition the higher domestic price encourages extra production, thereby increasing self-sufficiency.

This was the policy which regulated British agriculture before accession to the EEC, and it has now been incorporated in the Community's sheepmeat and beef support schemes. The great political advantage of the system in Britain was that consumers were for years cushioned against the real cost of food, and largely ignorant of the way in which general taxation supported the agricultural industry. In Europe by contrast consumers have always paid higher food prices, directly supporting farm incomes in the food price. In the present EEC system they support them in the food price *and* in general taxation, but the relative weakness of the consumer lobby on the Continent, combined with the expectation that food should absorb a higher proportion of income than British consumers have been conditioned to expect, has not provoked a general protest.

Structural reform of agriculture

The principal tool in most countries' agricultural policy remains price manipulation, but this is a short-term response to the ongoing problem of low relative farm incomes. The only long-term solution which has been attempted is structural change of the industry, designed to ensure that farmers eventually receive an adequate income without government intervention. In practice this means reducing the number of farmers who share the income of the agricultural sector. In theory at least, providing out-migration were achieved at the appropriate rate, the incomes of those left in farming would remain on a par with wages in other sectors of the economy. Several countries have policies whose aim is to encourage this out-migration, but all have been small in scale and limited in their effect.

The methods employed to achieve out-migration fall into two categories, each designed to achieve a gradual increase in farm size. The first set of policies aims to discourage young entrants to the industry by providing alternative job opportunities, and in some countries this is necessary not so much to increase farm sizes as to prevent further fragmentation of farms through inheritance. Even without government intervention there has been a steady outflow of young people from the farming community to urban areas as a result of industrial development and the increased access to education and training which farm children now enjoy in developed countries. In some areas the outflow has indeed been such that rural communities have lost nearly all their young people, and the problem is now to halt the decline of the rural infrastructure and isolation of an ageing population. In these areas the economic case for restructuring agriculture tends to be subordinated to the social problems, and the continuing support of farm incomes is judged a reasonable price to pay for rural social stability.

The second group of policies aims actively to encourage existing farmers to leave the industry. Again this is happening naturally as the farming population ages and in many countries is not replaced by an equal number of young incomers. Government policy therefore aims generally to encourage early retirement, which makes farm amalgamation easier. This involves a payment for the farms and the provision of retirement pensions, which in many countries were unavailable to farmers until recently. The policy usually goes hand in hand with the effort to persuade young farmers to seek other employment, but neither policy has had much effect. One plausible reason is that many small farms are in fact part-time ventures which do not provide their owners' only income, in which case financial incentives to quit the industry are unlikely to be very effective. In 1980 half of Britain's 243,500 farms produced 90 per cent of farm output, which means that there was a large number of very small farms with a low output. Nearly 100,000 holdings were in fact smaller than 20 ha, and 120,000 farms were estimated to be less than full-time businesses.

Income supplements for farming families

Part-time farming is even commoner in Europe than it is in Britain, and it is itself a structural factor which could be exploited to solve the farm income problem and overcome the worst problems caused by rural depopulation. In many countries a high proportion of small farmers already have two jobs (90 per cent of Japan's 5 million 'farmers' are actually employed part-time in industry). For part of the day or week they work on the farm, spending the rest of their time in local factories where special shifts are often operated for

Table 3.3. Farm structure expressed in terms of size and volume of business in the EEC and some member countries, 1979 (Based on data in *The agricultural situation in the Community*, EEC Commission, 1980 Report)

Size	EEC* 9		United Kingdom		France		Germany		Italy*	
	No. of farms (%)	Total ha (%)	No. of farms (%)	Total ha (%)	No. of farms (%)	Total ha (%)	No. of farms (%)	Total ha (%)	No. of farms (%)	Total ha (%)
Hectares										
1–5	41.9	6.2	14.5	0.6	19.4	1.8	32.1	5.5	68.2	21.4
5–10	17.4	7.3	12.6	1.4	14.9	4.1	18.9	9.1	17.5	15.9
10–20	17.6	14.7	15.5	3.4	21.2	11.9	23.1	22.2	8.4	14.9
20–50	16.8	30.0	26.2	13.1	31.0	37.8	22.1	43.8	4.1	15.7
Over 50	6.3	41.9	31.3	81.4	13.5	44.4	3.7	19.4	1.8	32.0
European size unit†										
Under 2	44.3	9.0	24.1	4.5	26.6	3.6	27.1	5.9	65.9	21.5
2–4	16.0	8.6	14.0	4.8	13.7	6.2	17.2	7.5	17.1	16.8
4–8	15.2	15.3	17.2	10.6	20.8	15.9	20.3	16.3	9.8	17.6
8–16	14.0	24.4	18.6	17.5	23.6	31.7	21.7	31.7	4.5	14.7
16–40	8.7	27.0	17.9	30.3	12.8	30.3	12.2	31.1	2.0	13.5
Over 40	1.8	15.7	8.2	32.3	2.5	12.3	1.5	8.5	0.7	15.9

* 1975
† This measures the size of the farm business. Small farms are classed as less than 2 ESU, whose income = approximately the return from 6 ha of wheat. Medium farms are from 2–8 ESU, while large farms are over 8 ESU, and their income is approximately equivalent to the return from 22 ha of wheat.

their benefit. In some countries transport is even provided by companies to take farmers up to 100 km to work. Since the farm in this case does not need to provide the total income, producer prices could be set high enough to guarantee sufficient production to meet demand, at a level of income determined in relation to the optimum-size producer rather than *all* producers. This policy could be combined with direct income supports for farmers whose income did not reach the minimum level, in the same way that welfare payments are already made to other low-income families. This has the virtue of providing reasonable incomes without raising consumer prices artificially. The procedure could also be arranged so that such payments continued only for the lifetime of the incumbent farmer and his wife, so that farm amalgamation was encouraged, or at least some limit set to the state's liability. Such payments are strongly resisted by farmers' organizations, however, which understandably prefer to see farm prices maintained and farmers' self-respect not harmed by the explicit payment of income supplements.

It must be stressed that there is unlikely to be a definitive solution to the problem of low relative farm incomes, though it is possible that the problem could become less serious if incomes in the rest of the economy cease to rise as fast as they have in the past. Research and development have constantly increased agricultural output, and with it downward pressure on incomes. Technology also increases productivity, and the minimum size of the efficient farm is constantly increasing. If incomes are not to fall, the number of farms must therefore continue to decline, so out-migration from the industry must continue unless governments decide to maintain the farm population at an artificially high level for social reasons. This detail is often overlooked, and the impression exists that once the existing large number of small farmers have left the industry, subsequent natural out-migration would alone sustain the farm population at an optimum level, so government intervention could cease. The likelihood is, on the contrary, that some form of intervention will continue to be necessary, even if structural reform is achieved in the long term.

POLICY AND POLITICS

All government policy is subject to opposition by organized interest groups, and none more so than agricultural policy. These groups include foreign governments and organizations as well as domestic interests which range from consumers and conservationists to producers' organizations. In all countries producers' organizations campaign publicly and lobby privately to protect farmers' interests. They are consulted formally and informally by governments on prices, import policy and the desirability or feasibility of self-sufficiency, on export potential, and all kinds of assistance which will ensure farmers' cooperation in the achievement of government policy. In France the farming vote was still important enough in 1981 to influence the presidential election campaign, and in the USA the three largest dairy cooperatives alone allegedly spent $1.2 million to sponsor candidates in the 1980 elections. Commodity groups within the farming industry (cereal growers, beef producers, dairymen) further complicate the issues, just as different trades unions diverge in defense of their particular interests. Most countries also possess unions of agricultural workers which exist to improve working conditions and their members' wages, and therefore tend to support their employers' arguments for higher prices and a strong agricultural industry, though they naturally want a greater share of the resulting benefits for their members. The farming industry may also be supported on major issues by the food industry, which is a major customer for many agricultural products, though again the two groups will be in competition on other matters.

Interest groups which oppose agriculture include consumer organizations and animal welfare agencies, conservationists and even quasi-government bodies like the British Nature Conservancy Council, which is often in conflict with the Ministry of Agriculture. Another factor which recently complicated the collision of interests was the formation of ecology or 'green' political parties, whose one unifying policy is opposition to the destruction of the environment, which they accuse agriculture of accelerating.

With this exception, political parties in the west have rarely been interested in agriculture. In the Communist world this is not so, of course, since agricultural policy there is wholly integrated in an ideology of state control and collectivization as the ultimate ideal. In the west it is generally assumed that farmers' politics incline to right of centre, but this is not a safe assumption, for in many countries there is a left-of-centre agricultural vote which complicates policy making. Party politics as such rarely enter into agricultural policy, however, since most parties draw most of their support from the urban majority, and the farming vote is declining even in countries like France where it long sustained a policy favourable to agriculture. In times of rising prices parties which rely heavily on the urban vote may raise the issue of farm prices and farm support, and propose modifications intended to reduce consumer prices. This is normally as far as most parties go, although there is one topic which perennially unites the farming population the world over when it becomes a political issue – land ownership. In developing countries, land reform policies figure prominently in party programmes and cause some of the most heated debates in election campaigns. In developed countries the issue barely interests the general public, but can be relied on to rally left-wing opinion and unite farmers solidly in opposition to any suggestion of land nationalization. The only other issue which raised equal passions in the 1970s was the acquisition of land by foreign interests, and governments in Europe and North America were lobbied to introduce obstacles to what was invariably represented as a threat to national food security.

If agricultural policy making is complicated at national level, at international level the complexity can become alarming. Governments must constantly negotiate trade agreements, tariff barrier reductions, access agreements, etc., which hinge invariably on the argument that agricultural produce can be allowed into one country only if it is itself granted a reciprocal right to export industrial products. Where imported farm produce is in direct competition with domestic produce, governments come under intense pressure to please their own electors at the expense of foreign trading partners and the world market, and the temptation is enormous to trade exclusively with 'friends', even colluding to deny access to produce from other sources. As the Brandt Commission has insisted, however, the temptation will have to be resisted if the resulting inequities are not to lead to long-term resentment and even conflict between the rich and the poor countries of the world.

THE COMMON AGRICULTURAL POLICY OF THE EEC

The complex nature of agricultural policy has probably never been more clearly illustrated than by the Common Agricultural Policy (CAP) of the

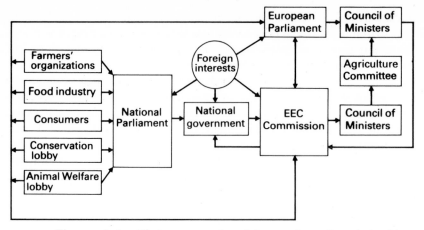

Fig. 3.2 A simplified representation of the complex policy relationships involved in the determination of EEC agricultural policy

EEC. Its complexities defy analysis in the scope of this chapter, but since it combines all the intervention methods outlined earlier, and has been subjected to most of the criticisms, a summary of the main issues will provide a concrete example of the problems which arise from the effort to reconcile multiple objectives – national, European and international – in a coherent agricultural policy.

The objectives of the CAP, as defined in Article 39 of the Treaty of Rome which established the Community, are:

(a) to increase agricultural productivity by promoting technical progress and by ensuring the rational development of agricultural production and the optimum utilization of all factors of production, in particular labour;
(b) thus to ensure a fair standard of living for the agricultural community, in particular by increasing the individual earnings of persons engaged in agriculture;
(c) to stabilize markets;
(d) to provide certainty of supplies;
(e) to ensure supplies to consumers at reasonable prices.

Compare this with the objectives of the 1947 Agriculture Act which regulated British agriculture before accession to the EEC, which were:

to promote a stable and efficient industry capable of producing such part of the nation's food supply as in the national interest it is desirable to produce in the United Kingdom, and to produce it at minimum prices consistent with proper remuneration and living conditions for farmers and workers in agriculture and with an adequate return on capital invested.

The policy objectives are very similar, though the difference might be noted between the 1947 Act's reference to 'minimum prices' and the Rome Treaty's reference to 'reasonable (consumer) prices'. The means of achieving the objectives differ markedly, however. British policy relied on a deficiency payment scheme which guaranteed minimum producer prices while simultaneously maintaining low consumer prices. The EEC policy by contrast relies on import control, using a combination of variable import levies and support-buying operations known as *intervention buying*. Import levies ensure most of the time that domestic producer prices are not forced down by outside competition, and when domestic oversupply threatens to depress prices the excess produce is purchased and stored in intervention stores. The excess is then sold on the world market, and if the sale price is lower than the purchase price, the difference is made up from the EEC's general budget (Fig. 3.3).

This policy unquestionably maintains farm incomes, but it is inequitable to EEC consumers, who pay higher food prices than those paid by buyers on the world market who purchase the subsidized surpluses of produce taken into intervention. In defense of the system it must be said that part of these surpluses are distributed through the EEC's food aid programme, so that relatively well-fed and wealthy Europeans are subsidizing the Third World's undernourished population. This apparent generosity is nonetheless counter-balanced by the fact that the EEC import levy system operates against the interests of Third World countries by denying them a market for their agricultural produce.

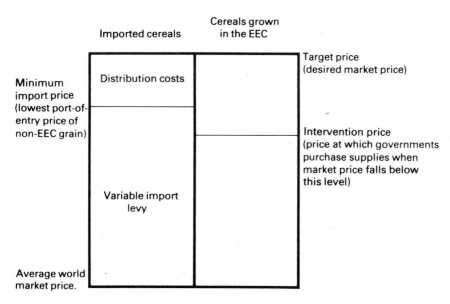

Fig. 3.3 A schematic representation of the relationships between the various prices used in the EEC cereal regime

The import levy system also operates against developed countries which export food to Europe, and the risk of retaliatory controls on EEC exports to them is always a possibility. Much more serious, however, is the case of Third World exporters, who often have no alternative market, and insufficient political or economic weight to threaten effective retaliation. Some developing countries signed an agreement with the EEC (the Lomé Convention) which allows them to export agreed quantities of agreed commodities to Europe without paying levies, but for some commodities these quantities are too low to sustain the signatory countries' previous level of exports. A good example is sugar, one of the main export earners of several Lomé countries, which is also one of the most profitable arable crops grown by European farmers. Since the foundation of the EEC, sugar beet production has risen to the point at which the Community as a whole is self-sufficient in sugar (though individual member countries may not be). Imports of sugar from cane producers in the Third World are therefore unnecessary, and EEC self-sufficiency has been achieved at their expense. It is also argued that the increase in sugar beet production in Europe has involved a mis-allocation of scarce resources which could be used to grow alternative crops which are not in surplus, and the substitution of high-priced beet sugar in place of lower-priced cane sugar has not benefited European consumers. The policy's effects have even extended to the refining industry, since factories designed to process cane sugar cannot handle sugar beet, so there have been plant closures resulting in company losses and redundancies.

The original EEC policy of import levies and intervention buying has been modified for certain commodities, along the lines of the deficiency payment scheme operated in Britain before accession to the Community. British governments rejected intervention buying because consumers pay higher food prices and, as taxpayers, subsidize losses incurred in selling surpluses on the world market. The EEC made some concession to this view when it agreed to sell some intervention stocks of beef and butter at subsidized prices to special consumer groups, mainly retired and low-income families. Another concession in 1980 allowed for deficiency payments to British beef and sheep producers, the hope being that this hidden subsidy would prevent a further decline in British meat consumption by not passing its cost on to consumers directly in the meat price. In the rest of the Community, however, intervention remains the major policy instrument.

Any price support scheme benefits large farmers who can change their systems in response to price incentives. However, small farmers and others whose options are limited by adverse climatic or topographical conditions cannot generally benefit from such incentives, so further support is needed which is not related to price structures. The EEC has such schemes which aid farmers in 'less favoured areas', and these schemes are effectively social measures designed to maintain a rural population. With this exception, the structural part of the CAP has had little effect in reorganizing European agriculture, the main reason being that price support absorbs most of the

CAP budget, leaving virtually no funds for long-term programmes. A structural plan (the Mansholt plan) has existed since the 1960s, but has never been implemented. Changes in the industry's structure are nevertheless taking place throughout Europe under the kinds of pressure identified earlier. Farm sizes are increasing, the number of farmers is declining, but not fast enough to improve the level of farm incomes. The principal obstacle to formal encouragement of this natural trend is the conviction of many EEC governments that the rural population must be maintained, and this social objective overrules arguments for a smaller number of larger farm units. To support small farmers on uneconomic units EEC governments have created job opportunities for part-time farmers, but commodity prices continue to be determined in a manner which reflects the income needs of these uneconomic holdings, and this gives even greater benefits to larger, efficient units, widening the gap within the farming community.

The CAP has without doubt succeeded in improving the standard of living of the agricultural community. It has also increased the EEC's capacity to feed itself. However, the question which is increasingly asked is at what cost? The policy has consistently absorbed over 70 per cent of the EEC budget, and since this allocation has gone on short-term price support rather than structural reform, the commitment is an open-ended one. With the accession of Greece in 1980 and the projected accession of Spain and Portugal in the 1980s, surpluses will exist not just in temperate products like milk and beef, but also in Mediterranean products like oranges, wine and olive oil. This will create more tensions within the Community, and almost certainly more pressure for reform of the CAP. Whatever changes are made, one outcome is certain: they will be made by politicians under pressure from the different interest groups which make up their constituency. The changes will therefore be the result of many compromises, and no single group will achieve all it seeks to achieve. Foremost among those lobbying the politicians will be the farmers' representative organizations, which may find their political influence waning in countries like Germany and France where it has traditionally been strong. However, those who hope that the Community will eventually downgrade the priority of agriculture's claims overlook both the strength of the farming lobby on the Continent, and the fact that the CAP *is* the only common policy to which the Community clings as a symbol of unity.

4

Resources and opportunities

Like any production process, the management of the farm business involves the efficient use of resources and the exploitation of every business opportunity. The principal resources of agriculture are land, capital and labour. These are complementary to each other, and within limits may be substituted for each other. Capital in the form of machinery may be substituted for labour, and where labour is at a premium this indeed becomes necessary. Where there is a surplus of labour, however, as there is in many developing countries, the substitution of capital for labour may be socially harmful as well as wasteful.

As important as land, capital and labour in determining the farming system is the existence of a broad industrial base and well-established infrastructure of communications, transport, water and power supplies, and even education, without which the progress of agriculture is restricted. Modern agriculture is a highly sophisticated industry, as a glance at the farming press or a visit to an agricultural show readily confirms. It uses a formidable range of complicated techniques which demand more than superficial understanding, machinery whose use and maintenance demand skilled attention, and chemical herbicides and pesticides which need careful handling. These have to be produced by advanced engineering and chemical industries which carry out research and develop techniques for implementation on the farm. The existence of these industries, like the existence of a literate, educated farming community, is an opportunity we take for granted in the west, but their absence in developing countries is a severe constraint on agricultural development.

The other major advantage which is taken for granted in developed countries is the existence of efficient marketing channels to distribute farm produce. In all but subsistence agriculture, the market is at least as important as land, capital and labour in determining farm systems, and it is the absence of a developed marketing system and the failure to expand marketing opportunities which prevent many developing countries from enjoying the full benefits of the green revolution which is transforming their agriculture. After all, if the twin functions of agriculture are to provide food for consumers and income for farmers, the marketplace provides the commercial opportunities without which none of the resources would have much economic value. Before considering these resources, a preliminary note on the market for agricultural produce is therefore needed, which a later chapter will expand.

THE MARKET

In deciding which enterprises to incorporate in his farm system, the farmer has to consider the market potential for each. He tries to assess what quantity and quality of product the market will take at various price levels, the relative merits of the different outlets, and the likely return from each. The market opportunities open to farmers, and therefore the complexity of the marketing decision, have become much greater with the growth of the food processing industry. Very little farm produce in the developed world reaches the consumer in its raw state. Three-quarters of the food eaten in Britain and Germany in 1980 had been processed in some form. Some was simply graded and packed, but much of it underwent processing which completely transformed it – e.g. fruit conserved as jam, or pigs turned into sausages.

The existence of a large, sophisticated food industry has created totally new opportunities for farmers, whose systems have sometimes been radically altered as a result. The frozen foods industry introduced brand new crops as well as making vegetable production feasible away from large centres of urban population where it had previously been concentrated. The industry sponsored the development of calabrese, for instance, and pioneered research into the husbandry techniques in collaboration with farmers before introducing the vegetable to potential growers. It even developed harvesting machinery so that it could present farmers with a complete package of proven measures which reduced the risk attached to any new farm enterprise.

Other sectors of the food industry introduced cash crops which can be integrated into existing arable rotations. They include lupins, grown for animal feedingstuffs, durum wheat for pasta making, and sunflowers and oilseed rape for the growing vegetable-oil market. Oilseed rape rapidly became a very profitable break crop on British farms (Ch. 6), and in the USA cropping patterns were revolutionized by the introduction of soyabean, which now supports a huge processing industry producing protein for a growing range of meat-substitute products. The stimulus to grow oilseed and soyabean both resulted from the industry's need to find alternatives to cholesterol-high animal products which had been associated with health risks, and their impact on farming systems demonstrates how agriculture is increasingly shaped by the food industry, of which it is still the major supplier of raw materials. This applies equally to developing countries, where agriculture-based industries are often the first to be developed because their basic input is already available. The problem there is that non-food crops like cotton and jute have frequently been developed to support fibre-processing rather than food industries, and this can seriously reduce the capacity of existing agricultural resources to feed the indigenous population. Multinational companies which have built processing plants which encourage non-food crops in the Third World have therefore been criticized, but their influence in the west has generally been less open to question.

The food industry is highly market-oriented, and tries to produce new products of value to consumers as well as ensuring that all products are in the best possible condition. Great emphasis is therefore placed on the quality of raw materials, and this affects farming standards very considerably since quality payments can significantly increase revenue. To ensure the highest possible quality for freezing, dehydration and canning, food companies generally offer farmers written contracts to produce a given volume of produce of a specified quality, for which they undertake to pay a specified price. This reduces their risk of not obtaining sufficient supplies of high-quality produce, and the farmer's risk of not finding a buyer, but the arrangement is not without its problems. It removes the firm's option to shop around for the right produce at a lower price, and the farmer's freedom to seek higher prices on the open market. In both cases this has led to the integrated production and sale of some farm produce (*vertical integration*), either by companies which own farms (especially pig and poultry units) and sell through their own outlets, or increasingly by farmers who sell direct to the public through farm shops. More commonly, however, production and processing or packing are integrated, and selling is done through a conventional retailer.

Vertical integration has always been viewed with suspicion by the agricultural industry, and many farmers are also suspicious of contractual agreements with food companies, believing they will lead to a loss of independence. There is nonetheless little evidence that farmers have been turned into mere employees. Nor does the food industry generally wish to become its own supplier, if only because the return on the investment in farming is so low compared with alternative investments it could make. Whatever the impact of the industry on farm systems, it has therefore had relatively little effect in changing the markets traditionally open to farmers, increasing the options rather than diminishing them, as Chapter 12 will argue.

LAND

The amount of land devoted to agriculture in many developed countries is declining as the result of progressive urbanization. At the end of the nineteenth century 87 per cent of the total area of the UK was farmed, compared with 79 per cent in 1980. In the same year another 7.5 per cent was allocated to forestry. These figures compare with 55 per cent of the total area of the USA which is devoted to agriculture, and 24 per cent to forestry. A significant proportion of the remaining land area of the USA is exclusively reserved for recreation, but in the UK and most of Europe very little land is devoted exclusively to recreation or even nature conservation, and national parks and nature reserves frequently double as working farms. In 1980, 21 per cent of the land area of England and Wales was designated as National Parks, 'sites of special scientific interest' or 'areas of outstanding natural beauty'. This often leads to conflicts of interest between farmers who own and work this land and the general public who see it as a national asset over which they can walk and picnic at will. At the very least, it means that agricultural land is never truly private in the way that a manufacturer's plant is private, and few industries are so subject as a result to constant public scrutiny.

The amount of land devoted to agriculture is influenced by the prosperity of the industry itself, and the demand for land for other uses. In most industrialized countries it is the latter factor which assumes greater importance, the prosperity of agriculture having more influence in determining the actual use to which land is put. Although the area farmed is declining in many countries (between 1971 and 1978 the average loss was 50,000 ha/ annum in the UK), the productive potential of much of it is increasing as the result of technical improvements. Even so, productive potential is still limited by physical factors like soil quality, climate and relief (altitude and aspect), which impose long-term restrictions on the range of crops that can be grown, the level of yield to be expected, the regularity with which such yields may be expected, and the cost of obtaining them. On the basis of these factors a land classification has been devised in the UK which is a use-

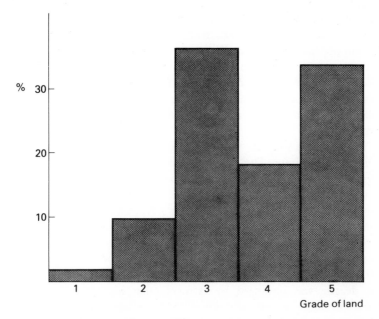

Fig. 4.1 Agricultural land classification of the UK, showing the relative proportions of the different grades

ful indication of the productive potential of the five grades of land identified, though the economic value of farms depends also on their geographic location, chiefly proximity to markets.

One object of this classification was to direct the development of urban areas or forestry away from the most productive land, but as a planning tool its use is limited. In a particular small area it may be possible to steer development away from land of the top two grades, but at the macrolevel the problem is that most of the lowest grade land is in areas geographically remote from centres where development is likely to take place, while some of the best land is immediately adjacent to growing urban centres.

If the amount of land available for agriculture changes over time, so too does the size distribution of farms and the ownership pattern, both of which influence the farming systems practised. Some systems, for example, can only be carried out successfully on an extensive scale, while others are better suited to small holdings. The size structure also affects the type of agricultural policy adopted, and the policy will in turn affect the size structure and the ownership pattern of farms.

In most developed countries farm size has increased throughout the twentieth century. This has been partly the result of improved technology and mechanization, themselves in part attributable to a decline in the agricultural labour force. Between 1900 and 1980 the number of holdings in the UK (defined in fairly comparable terms) fell by about 175,000 to 243,500, and the average farm size rose from 25 to 50 ha. In 1980, farms over 100 ha occupied 50 per cent of agricultural land, an increase of 10 points over the preceding decade. As long as the object was to produce as much food as possible, a large number of relatively small units was desirable, because production per hectare is generally greater on small farms. However, when production at any price is no longer the major policy objective, but least-cost production becomes important, larger farms have an advantage because they can benefit from economies of size which reduce production costs. Producer prices can thus be allowed to fall without causing severe loss of income (Ch. 3).

The evidence suggests that the increase in farm size observed in the past is likely to continue. The economies of size which existed in the mid-1970s suggested that farm sizes of 40–100 ha (depending on enterprises) are the most efficient, and scarcity of resources and economic pressures are likely to reinforce this trend. In many countries government policies are actually designed to accelerate the trend, but the change takes place anyway simply because many farmers with spare capacity in terms of machinery or management purchase neighbouring parcels of land whenever possible. It has been estimated that 80 per cent of the land which changed hands in Britain in the last decade was absorbed by farmers enlarging their existing holdings. These farmers can generally afford to pay a higher price for the land than other prospective buyers precisely because they may not need to invest in other fixed resources to exploit it. Equally, they may not expect a full market return on the added investment, since they can spread the financial charges associated with the purchase (interest and repayment of loan) over the whole of their enlarged business, instead of having to meet these charges from the new land alone. They are thus not obliged to farm it as intensively – and perhaps not as well – as if they needed to make a reasonable living from it while simultaneously meeting the financial charges on the new holding.

The effect of farm amalgamation is generally considered beneficial to the national economy, but the fact that most farms are sold to existing farmers means that agriculture is becoming a closed industry to which newcomers find access difficult because they cannot raise sufficient capital to compete with established farmers. The alternative way of entering the industry used to be as a tenant. In the early twentieth century only 12 per cent of the UK's farms were owner-occupied, whereas the figure in 1980 was 66 per cent. Most farmers certainly prefer to own the land they work, but tenancy has its advantages, one of which is that it does offer newcomers a relatively low-cost way of entering the industry without taking on a heavy burden of debt. It can also offer invaluable training experience, and good landlords have traditionally offered small tenancies to young farmers with the intention of promoting them to larger holdings as their experience grows. However, tenancy is in decline in most countries, and in some, active policies encourage the sale of rented farms to their tenants. Such policies have not proved necessary in most European countries, where land ownership has been made increasingly unattractive by high levels of taxation on unearned income (including rents) and heavy transfer duties on inherited land. Consequently many landowners have started to farm land themselves, or sold it and reinvested the money elsewhere. This has reduced the opportunities for tenant farming, though where land has been bought by institutions like pension funds or private companies, some has remained available for tenancy. The purchase of large estates by these institutions has been criticized by farmers' organizations, but there is little evidence that they are likely to farm it less well than the other institutions which traditionally held large areas of land – notably the church and old-established universities. From the farmer's point of view, all that matters is that his landlord should leave him alone to manage the farm business, and in this respect the new institutional owners are no more likely to wish to interfere in the technical problems of management than traditional institutional landlords.

One form of tenure which is increasing is *sale and leaseback*. Many farmers who own their land are desperately short of capital to expand or even maintain their business, and one answer is to sell the land (usually to an institutional buyer) on the guaranteed understanding that they can continue to farm it for a specified period, generally their own lifetime, but possibly their son's lifetime too. If a wealth tax were introduced in Britain, this would unquestionably become a very attractive proposition for many farmers, giving them virtually all the security of owner-occupation while simultaneously raising capital to run the farm.

The same advantages are often claimed for land nationalization, which would turn all farmers into tenants of the state. The difficulty is that land nationalization would almost certainly be *imposed* on the farming community, which is a totally different situation from freely chosen and individually negotiated sale and leaseback arrangements. Farmers also suspect that they would receive insufficient or even no compensation for land from any government strong enough to introduce land nationalization, in which case they might find themselves tenants without receiving the working capital raised by a sale and leaseback agreement. Such an arrangement also has the advantage of being a legally binding agreement, and breaches of contract by either signatory can be settled in the independent courts. Even a liberal tenancy agreement introduced by government in parliament as part of a nationalization scheme could by contrast be repudiated by a future government, and it is uncertainties of this kind more than political ideology which unite farmers in opposition to nationalization. Exactly like a home owner-occupier, they feel securer if they own the land they work, despite the improvements in tenancy agreements introduced in recent years. Owner-occupation offers them independence and flexibility of management which they fear would be eroded as a consequence of land nationalization. Advocates of nationalization usually argue that the state would be no worse a landlord than other institutions, but wherever land has been taken into public ownership agriculture has become increasingly dependent on a central bureaucracy which interferes with free management decisions, and ultimately decides farm policy in every detail for implementation by farmers-turned-employees. This has proved a real obstacle to agricultural self-sufficiency in an industry where rapid decision making is essential, since delay can lose a crop to bad weather or disease, or miss opportunities which result in under-production. The Chinese government admitted as much in 1980, and Russia and Rumania also acknowledged in 1981 that state-owned farms are far less efficient than privately owned plots, which in Russia produce about 25 per cent of all agricultural output on only 3 per cent of the land. This reluctant admission simply confirms that farmers everywhere work better when they are their own masters, and this, more than the strict ownership of land, is the source of the agricultural industry's resistance to land nationalization.

CAPITAL

Land is clearly a form of capital, in the sense that it is transformable into capital by sale, and it needs to generate income for its owner equal to that which he could achieve by alternative investments. In farm accounting, land is conventionally called *landlord's capital*, along with buildings and plant, even though the land is owner-occupied. Other forms of capital, including machinery and stock, are conventionally described as *tenant's capital*, which is distinguishable from landlord's capital in that it is *mobile*, i.e. it can be transferred from farm to farm by its owner. Tenant's capital can be a wasting asset, which means that it suffers from depreciation (loses value) as the result of wear and tear or obsolescence. This is rarely true of land, though it is true of buildings. A sub-category of tenant's capital is *working capital*, which is money tied up in stocks of fertilizer or crops growing in the ground, or goods sold but not yet paid for.

Where a farm is tenanted the farmer is responsible for finding tenant's capital, but the owner-occupier also has to find landlord's capital. In either case money may be borrowed from banks, insurance companies or government-supported bodies like the British Agricultural Mortgage Corporation, which finances farm purchase with capital contributed by banks, the Bank of England and the Treasury. Money borrowed for land purchase is invariably tied to a mortgage, requiring regular repayments of capital and interest charges which must be met out of income, the land providing security for the loan. Tenant's capital is more difficult to borrow where there is no land to offer as security, so real tenants can generally borrow less than owner-occupiers. However, in most countries except Britain special agricultural credit banks exist (often cooperative banks founded by farmers but generally state-supported) which offer loans on reduced terms to their shareholder borrowers.

The size of bank lending to agriculture is very substantial in many countries. In California, the largest of the USA's farming states, farm debt increased 180 per cent between 1970 and 1979. In Britain in February 1980, more capital was on loan to agriculture than to any other industry. The figures also showed a very rapid expansion in borrowing, which had risen by 250 per cent in 5 years. The cost of this borrowing had doubled in the same period, to 5.4 per cent of agriculture's gross output in 1979, but as Table 4.1 shows, the proportion of borrowed capital to asset value has actually declined, mainly because of rising land prices.

Most changes in a farm system require capital, and it is often lack of capital which prevents worthwhile changes from being made. For instance, intensification of livestock production means that more stock must be purchased, or some animals retained which would otherwise be sold, thus reducing revenue. It also generally means that more buildings and fixed equipment are required to house the stock and its winter feed, and additional capital may also be needed to improve the land on which the extra stock are to be kept. This may involve expensive capital works like drainage, or less costly improvements like the ploughing up and reseeding of rough grassland. In most countries governments have encouraged such investments by providing generous grants or subsidies, or by subsidizing interest rates on borrowed capital. The motivation may be to increase production, or to increase productivity, and in the latter case it may be a social subsidy designed to improve small farm incomes rather than a move to im-

Table 4.1. Estimated balance sheet for British agriculture 1970–80
(£ millions)
(Based on data supplied by the Agricultural Mortgage Corporation)

	1970	1974	1980
Assets			
Land	5,800	16,870	37,000
Machinery	800	1,170	4,200
Livestock	1,300	1,780	4,500
Crops and stores	570	900	2,000
Debtors	200	320	700
Cash	200	340	500
	8,870	21,380	48,900
External finance			
Mortgages	200	290	455
Merchant credit	250	300	850
Miscellaneous	20	20	90
Private loans	350	300	350
Banks	500	870	3,150
	1,320	1,780	4,895
External finance as % assets	14.9	8.3	10.0

prove the actual efficiency of the industry. However, grants for long-term capital works like drainage and building are a means of improving the whole capital base of the industry which most governments have judged worthwhile.

TECHNOLOGY

Technological change has been a major stimulus to change in agriculture as it has in other industries. The stimulus has usually sprung from within the industry, as in the case of inventions made by farmers to meet particular problems, but occasionally developments in totally unrelated fields have had unforeseen applications in agriculture. Most research is funded by the industry in collaboration with governments and agro-industries, and it is this partnership with commercial companies which has been criticized for concentrating research on the development of techniques which expand agriculture's reliance on their products. The research effort dedicated to the mechanization of agriculture has for example been attributed to the machinery manufacturers' desire to expand their market. Similarly, the emphasis on research into high-yielding plants which respond to high fertilizer applications and chemical treatment is attributed to a conspiracy among chemical manufacturers to boost their sales. It would be more realistic to attribute both to the consistent drive by governments to expand agricultural

Fig. 4.2 Mechanical planting of brussels sprouts. The machine on the right plants 10,000 plants/ha, while the machine on the left applies powdered insecticide against cabbage root fly. The machinery achieves higher accuracy in planting with much less labour than hand-planting (Farmers Weekly)

production and increase productivity in a century of rapidly expanding population, insufficient food, and political and economic pressures to increase agricultural self-sufficiency.

Efforts to improve agricultural techniques are constantly being made with the object of increasing saleable yields, replacing scarce resources and introducing new products. The saleable yield can be raised directly by breeding plant and animal varieties superior in quality, in resisting pests and diseases, and in converting inputs into usable output. Yield can also be raised indirectly by more effective control of pests and diseases during the production and storage of produce, and by improved husbandry techniques which reduce losses and ensure the efficient use of fertilizer, irrigation water, energy, etc. Research has not just concentrated on the drive to increase saleable yields, however. It has increasingly been directed towards the replacement of scarce resources, chief of which in the developed world is human labour. Research has therefore concentrated on the development of machinery capable of doing the laborious farm work which men no longer wish to do. The agricultural workforce has declined steadily over the last two centuries in the west, and the unavailability of labour, and its high cost when it is available, have been a strong stimulus to research into mechanization. Mechanization in its turn has brought some of the greatest changes to agriculture, some of them not immediately obvious to the layman. The introduction of the first tractors, for example, not only allowed work to be done faster (since tractors are quicker than horses); it also reduced the time spent in looking after horses, their tackle and their feed. More important, it significantly increased the land available for human food production, since thousands of hectares no longer had to be allocated to producing feed for working horses. Today the emphasis is on devising means of reducing the energy requirement of the mechanical workhorses which depend on diminishing reserves of fuel oil.

The introduction of new crops by the food industry, or new uses for existing crops, has already been mentioned as a rich source of new opportunities for farmers which simultaneously enlarges the range of consumer choice. The impact of new crops is even greater where areas previously unsuitable for cultivation can be brought into production – the development of drought-resistant plant varieties being one example. Even the introduction of existing crops to a new area, perhaps after modification by plant breeders and the development of new husbandry techniques, can significantly affect the farming systems of developed and developing countries.

One consequence of technical advance is the need to study how new inputs or techniques are best incorporated into existing farm systems, and how existing systems need to be modified to derive the greatest benefit from them. How should a new crop be fitted into rotations, and what will be the effect on the other crops in the rotation of the pests and diseases associated with the new crop? How will a new harvester affect the method of growing a crop, and the labour organization at harvest time? Before any innovation can be widely introduced these questions must be answered, and though the basic research may be carried out by the original innovator, the consequences for whole systems have to be tested under farm conditions. This is one of the main functions of experimental and demonstration farms and advisory services working in conjunction with selected producers. The final test is nevertheless how well new components fit into individual farm systems, and only if farmers succeed in adapting them to their individual circumstances can the new technology be said to have proved itself. It is at this final stage that much promising research founders because it fails to work on a farm scale, or cannot be successfully integrated into farming systems.

LABOUR

The agricultural labour force is composed of farmers and their families and paid employees. Most of the latter are permanent full-time employees, but some are casual part-time workers employed at seasonal peaks. The most important fact about the agricultural labour force is that it is constantly declining. In 1851 agricultural workers represented 23 per cent of Britain's working population, but in 1980 the figure was 2.7 per cent. Though the proportion may still be higher elsewhere, the same long-term decline in the farming workforce is observed throughout the west. The relative proportion of farmers and family workers to paid employees is also changing. In the period 1960–80 the number of farm workers in Britain declined by approx-

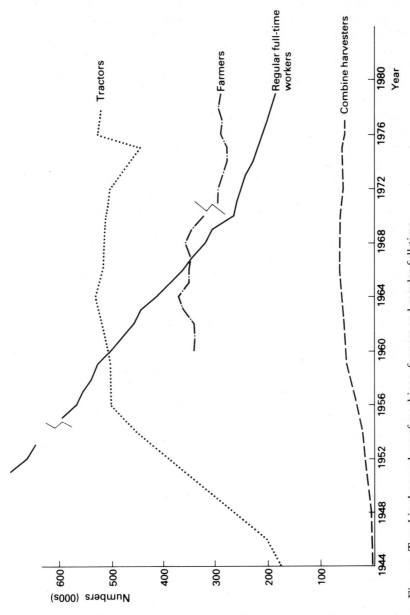

Fig. 4.3 Trend in the numbers of machines, farmers and regular full-time farm workers in the UK 1944–79 (Based on data in *Agricultural statistics*, HMSO, various years)

imately a third, and there are now 1.5 workers to every farmer, whereas the figure in 1960 was 2.0. As Chapter 3 noted, even some farmers are not fully employed on their own holdings, though this is less common in the UK than it is in Europe.

Despite the declining labour force, labour productivity in the developed countries continues to increase, partly as a result of the substitution of machinery for men. In Britain labour productivity in the farming industry increased by 4 per cent annually throughout the 1970s. Figure 4.3 shows the steady fall in the labour force compared with the growth in the number of tractors and combine harvesters on British farms. In the immediate post-war years there was a rapid increase in the number of machines on farms, but their number has not substantially increased since that time. Since output by contrast has constantly increased in the last thirty years, the *capacity* of machines on farms must have increased.

This substitution of more and larger machinery for men has been widely criticized for the effects which it has on the physical environment and for the high energy dependence of machines. In the face of recent high levels of unemployment it has been criticized as especially wasteful, the assumption being that substantial numbers of the unemployed could be redeployed in agriculture. The reality is that few people are now prepared to do agricultural work, and even fewer would be prepared to do so if they were expected to do without the machines which have taken some of the hard physical labour out of the job. It is even unlikelier that unemployed industrial workers would tolerate the low wages of agricultural workers, who have consistently been among the lowest paid workers in the economy (Table 4.2).

The shortage of labour causes particular problems for some farm enterprises and in some regions. The most labour-intensive farming systems are fruit-growing and horticulture, which also demand higher skills than many other farm jobs. Many crops like strawberries and raspberries still cannot be picked mechanically, and the casual workers who used to do this work have either disappeared or demand much higher wages. This has led to the growth of *pick-your-own* fruit, which many townspeople seem happy to do, though the solution is not helpful to a grower who lives far from centres of

Table 4.2. Agricultural workers' wages compared with other industrial earnings in 1979
(Based on data in *Annual abstract of statistics*, HMSO, 1981)

Full-time adult male	Weekly earnings (£)
Agricultural workers	75.21
Mining	109.55
Manual workers in manufacturing industries	97.90
Non-manual workers in manufacturing industries	117.70

population, and the damage and even vandalism caused by unskilled pickers has to be offset against the advantages of the system. For some crops like tree fruit, completely new growing techniques have had to be devised to overcome the labour problem. The normal fruit tree, with a canopy of branches high on a central stem, is difficult to pick. Single-stemmed forms have therefore been introduced which yield far fewer apples, but can be mechanically harvested. Since they can be planted much closer together than standard trees, the yield per hectare is still high, so a crop which was threatened by high labour costs is again economic thanks to a combination of scientific research, mechanization and the farmer's responsiveness to new techniques.

The causes of the declining agricultural workforce are multiple and complex. The consistently low wage level of farm workers has always encouraged men to leave the land for easier and better-paid jobs in towns. However, the continuing decline today is due more to the fact that men who retire are not replaced. Replacements are difficult to find because farming requires a wide range of skills of its paid employees, and few people have, or are prepared to acquire these skills. When labour *is* available it is also expensive, since skilled men expect higher wages than the traditional labourer. Consequently, one of the farmer's most important management tasks is to ensure that it is used with maximum efficiency.

The decline in the availability of labour has led to the growth of contracting firms which undertake farm work. Contractors have always existed to undertake specialist jobs requiring very expensive or highly specialized machinery which would be under-utilized on one farm. (Drainage work and crop-spraying by aircraft are good examples.) However, the increasing cost of all machinery and the lack of labour are causing more farmers to employ contractors to do work they would previously have done themselves. There are still specialist companies who do specialist jobs, of course, but in arable areas there are now firms which will carry out all farming operations – ploughing, sowing, spraying and harvesting, and this means that a farmer can run a 150–200 ha farm without employing regular workers, using contractors for virtually all the farm work. This is not yet common, but the escalating cost of machinery and labour is likely to make it more so in the future.

In this last case it is clear that the farmer's principal input to the operation of the farm is not his labour, but his management skill. All farmers contribute management skill and entrepreneurial judgement as well as labour, and traditionally these skills were acquired by experience, passed on from father to son. Today they are increasingly reinforced by college or university training which covers economic and management principles as well as the technical aspects of modern agriculture. It is becoming rare to find a young farmer who has received no formal training, and in some countries (e.g. Denmark and West Germany) no-one who has not received such formal training is allowed to run a farm.

Advances in technology make it imperative to up-date farming knowledge all the time, and in this respect the farming press plays a leading role in most developed countries, reinforced by television, radio and data retrieval systems, which provide market reports and technical information. In virtually all countries, however, the key role is played by advisory (or *extension*) services provided by governments and by private consultants. Through research, demonstrations on working farms or experimental stations, and through consultations with individual farmers, these services maintain and increase the industry's efficiency, ensuring that the latest proven techniques are introduced with the shortest possible delays and the least possible risk. These activities are almost bound to increase output, and thus to exert further downward pressure on producer prices, which is generally the one thing which governments wish to avoid since this involves them in greater price support. In principle it is thus arguable that advisory services offered by governments are working against government policy. In most countries, however, it is accepted that under normal circumstances the adviser's function is not to implement overall government policy, but to assist the individual farmer.

The farming industry and individual farmers are thus backed up by a range of supporting industries and services which help them to respond to existing resources and opportunities, and create new ones which maintain the industry's dynamic character. In the end it is nevertheless the farmer's own technical and management skills which represent his best resources, and – where they are deficient – his worst constraint. Education and access to advice can develop these skills, but they cannot replace the native capacity to make sound judgements and carry them into effect. Farming calls for a range of skills wider than most people realize, and it would be unreasonable to expect all farmers to demonstrate equal competence in exercising them. It is the object of the next chapter to outline some of these skills on which the success or failure of the farm business depends, before describing in the remainder of the book the particular farming systems in which these skills are put to use.

5

The management of the farm business

The term *agribusiness* symbolizes for many people all that they dislike about modern agriculture, but agriculture is obviously a business, and the farmer a manager whose responsibility exceeds that of the average industrial manager. Farm management has indeed been taught in universities and colleges far longer than business management, but the object of this chapter is less to examine the formal principles of the subject than to describe the decision-making processes farmers face as managers.

In this role the farmer performs functions which in other management situations are performed by several individuals or whole departments in large companies. He is responsible for overall policy and makes day-to-day decisions on the level of inputs, employment of labour, etc., as well as working on the job himself. He thus acts as company director, works manager, foreman and labourer. In purchasing and selling commodities he combines the functions of purchasing and sales executives, and he is also financial controller and accountant, overseeing the payment of accounts and managing farm debt, and ensuring that financial as well as physical targets are met. The one job he is very ill-advised to do himself is to be his own vet. The diseases of farm livestock and their treatment need the same specialist care as human illness, and though the farmer may administer treatment under a vet's guidance, he is foolish to try to diagnose and treat conditions himself.

The time allocated to these functions varies with the complexity of the farm business. On a small farm the farmer is concerned more with implementation and control than with repeated shifts of policy, but on large farms policy making and financial management absorb far more time. Few farmers escape responsibility for all of them at some time or another, however. The three main planning decisions are what to produce, and how much land and other fixed resources to commit to each enterprise; what yield to aim for and how to achieve it; and how the output should be marketed. These decisions are clearly interrelated. The enterprise decision is also a marketing decision, for instance, since it is useless to produce a commodity for which there is no market. The basis of all the decisions is a comparison of the projected revenue and projected cost of alternative enterprises, or alternative methods of producing a given yield, and though it is not the only criterion, profitability weighs heavily in the final choice.

THE ENTERPRISE DECISION

The enterprise decision rests primarily on profitability and the farmer's personal preferences. Profitability is affected by market factors, price, and production cost, all of which may be affected by government policy towards guaranteed price support and the subsidization of production inputs. Individual preferences influence the decision in two ways. First, the farmer may actively dislike certain enterprises – e.g. livestock, or he may have a particular enthusiasm, like breeding highly-productive dairy cows or prize sheep. Second, his conception of what constitutes an acceptable level of profit clearly enters the analysis. If his preferred group of enterprises (say, beef and sheep) does not yield sufficient income, he may have to introduce enterprises he dislikes which do. Alternatively, he may decide he values a higher income less than doing what he likes.

A farmer may decide to produce only one commodity, or a limited range of related commodities (wheat, or cereals, for example). Though this is common in industry it is rare in agriculture, however, so the farmer's most difficult task is to design a system in which the potential conflicts between enterprises are minimized and the beneficial interactions fully exploited. This management function is not particularly important in many industries, but it is critical to farm management, and agriculture was one of the first industries to use mathematical techniques of linear programming and simulation analysis to help solve the problem.

On most farms the enterprise decision is an infrequent one. The range of possible enterprises is limited by natural constraints, the availability of land, labour, capital, the farmer's experience and government policy. A shortage of land might point to intensive livestock production, but inexperi-

Fig. 5.1 A farmer using his mini-computer to check accounts

ence or dislike of the system could weigh against the choice. Shortage of capital might force a farmer to plough up the whole of a previously grass-and-arable farm for cereals, in order to generate a high return in the shortest possible time. A newcomer to farming might prefer an enterprise like milk production which brings in a regular income, but shortage of labour might prevent the adoption of this labour-intensive system. The previous history of the farm may also be a limitation in the short term. A cereal grower might wish to change to milk production, which has a higher gross margin. However, the existence on the farm of harvesting and storage facilities and the absence of expensive dairy equipment could make him choose a lower income rather than the cost of investing in the new enterprise.

A complete change from one enterprise to another requiring major investment is clearly more difficult than a marginal change which needs few extra resources. The change from milk to cereals is a major one, requiring totally different management skills as well as investment, but the change from one cereal crop to another requires neither massive investment nor new skills. A marginal change is also easier to make in that the farmer can generally arrive at fairly accurate costings for the proposed enterprise, based on his own experience in the related field or on figures known for his locality. In both cases the underlying principle is nevertheless the same: that each unit of any resource should be employed where it earns the highest return. This is expressed in terms of *gross margin* per hectare, since for most farmers land is the scarcest resource which must be fully exploited. (If labour is scarce, gross margin per manhour is the appropriate measure, and if water is the limiting resource, gross margin per centimetre of water is the measure.)

The gross margin for any enterprise is calculated as follows:

Gross margin = Gross output – Variable production costs

In agricultural business management, *variable costs* are those which can be allocated to a specific enterprise, and which vary approximately in direct proportion to the scale of that enterprise. *Fixed costs* are those which do not vary with the scale of the enterprise, and cannot readily be allocated to a particular enterprise. Variable costs include such items as fertilizer, seed, agrochemicals and casual labour. Fixed costs include rent (notionally calculated where land is owned), permanent labour and financial charges, and are ignored in calculating the gross margin. Machinery and its associated charges are also treated as fixed costs, even though it takes more fuel, for example, to combine a high-yielding field than one with a lower yield. The cost of ploughing in both cases is identical, so conventionally all machinery costs have been neglected in the gross margin calculation. Whether they can continue to be neglected as fuel costs soar remains questionable, however.

The different gross margins of alternative enterprises are the basis of the enterprise decision. The most reliable figures available to farmers are those calculated for their own farms, but where new enterprises are concerned figures for the locality are available from the advisory services which can be modified to suit individual circumstances. Since these figures are based on past experience, whereas farmers are planning for the future, the gross margins must also be modified to anticipate probable changes in costs and returns. It is for this reason that farmers are eternally preoccupied with government policy, and that agricultural prices are negotiated in advance, since if farmers do not know with reasonable certainty when planning the year's work what price they will get at harvest, the risk increases and production may decline.

The gross margin calculation is relatively simple where a farmer wishes to make a marginal change in his farm system. Supposing he wants to increase his income by increasing to the maximum the hectarage of his most profitable crop, the effect of the change can be calculated by means of a simple budget. From Table 5.1, which shows the gross margins of some arable enterprises, it is clear that every hectare of land which grows winter wheat instead of spring barley yields an extra income of £116. No extra machinery is required in this case since both crops require very similar treatment, but a change from spring barley to potatoes would need additional resources, resulting in increased fixed costs which must be built into the calculation. More sophisticated procedures are necessary to do this, since the gross margin calculation neglects the interactions between enterprises as well as changes in fixed costs. These interactions may make an economically desirable change operationally difficult, and to neglect them is unwise. It might be economically desirable to grow 100 per cent wheat instead of a mixture of wheat and barley, but this would concentrate the labour requirement which was previously spread over two crops with different cultivation and harvesting schedules. The substitution of oilseed rape for barley would have the opposite effect, since rape is harvested earlier than wheat, so the labour requirement would be better spread.

Major changes in farming systems must be considered in relation to natural constraints and the availability of fixed resources. Some of the farm may

Table 5.1. Specimen gross margins for selected arable enterprises (Based on data from *Report on farming in the eastern counties of England 1979/80*, Cambridge University, 1980)

Enterprise	1979/80 (£ per ha)	Range 1970–80* (£ per ha)
Potatoes	1,231	303–2,827
Sugar beet	738	276–772
Winter wheat	398	338–507
Winter barley	343	271–404
Spring barley	282	237–404
Oilseed rape	430	296–473
Field beans	312	146–443

* All items valued at 1979 input prices

be unsuitable for cultivation, or the farmer may need a crop rotation as a means of disease control (Ch. 6). There may also be institutional constraints, like quotas applied by governments to limit the total land area devoted to particular crops or enterprises. The amount of fixed resources besides land also limits the area which can be allocated to a particular enterprise. Potatoes require a labour input of 120 manhours/ha as against 19.5 manhours/ha for wheat, for example, so a farmer with 6,000 manhours available can devote them either to 50 ha of potatoes or 308 ha of wheat, or some combination of the two. Given the profitability of each crop he can then calculate the likely return from each crop combination.

This principle can be applied to any number of fixed resources, and the ultimate amount of land devoted to each enterprise is decided in relation to the combination which produces the highest revenue through the optimum use of the fixed resources. The method of finding the ideal solution is *linear programming*, which involves the use of a computer – to which some farmers already have access, while others call on the advisory services or consultants to solve the problem. It is impossible to explain the technique adequately here, but it can be illustrated in very simplified form, using a farm system where wheat and sugar beet are under consideration.

First, all the available resources are listed accurately, and how much of each is required per unit of output, together with the gross margins for the crops on the particular farm (Table 5.2). Any absolute constraints are also listed, e.g. the size of any production quota, or any factor fixed by the farmer (such as the decision never to have more than a third of his farm in wheat at any moment). All these limitations are expressed as a maximum area available for each crop. For example, working capital = £35,000 = 140 ha sugar beet or 219 ha of wheat, or some combination of the two.

The next stage is to compare the different use made of the main re-

Table 5.2.

Farm resources available		Wheat		Sugar beet	
		Resources required per hectare	Maximum hectarage	Resources required per hectare	Maximum hectarage
Land (ha)	190	1	190	1	190
Working capital (£)	35,000	160	219	250	140
Spring labour (h)	400	1	400	4	100
August labour (h)	450	2.5	180	0	∞
Sugar beet quota (ha)	90	0	∞	1	90
Gross margin £ per ha		375		580	

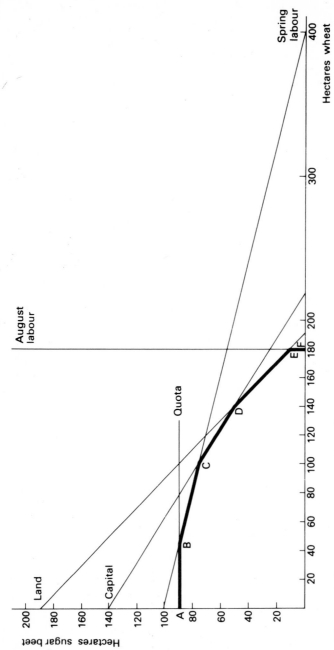

Fig. 5.2 A diagrammatic representation of the use of linear programming to determine an optimum crop mix

Table 5.3.

| | Resources | | | | |
	Land (ha)	Working capital (£)	Spring labour (h)	August labour (h)	Sugar beet quota (ha)
Available	190	35,000	400	450	90
Required by:					
Wheat	103	16,480	103	258	–
Sugar beet	74	18,500	296	0	74
Total	177	34,980	399	258	74
Unused	13	0	0	192	16

sources – land and labour – in alternative crop combinations. These are represented diagrammatically in Fig. 5.2, where the possible alternatives are shown on the *production frontier* ABCDEF. At point C, for instance, there are 74 ha of sugar beet and 103 ha of wheat, and the resource situation (shown in Table 5.3) is that 13 ha are unused because all the spring labour and working capital are fully employed.

All the points on (or inside) the production frontier represent feasible solutions. The problem is to decide which combination will yield the highest return. In order to calculate the gross margin for each possible combination, it is assumed that the fixed costs remain identical whatever the change. (The gross margin could not be used in this simplified manner to plan a real farm system *ab initio*, but the principle of the exercise would be the same.) At point B, where the farmer is using his total quota of land for sugar beet and his spring labour is fully employed, the margin is:

90 ha sugar beet at £580/ha	£52,200
40 ha wheat at £375/ha	£15,000
	67,200

The margins at the other points on the production frontier are given in Table 5.4, which shows that 51 ha of sugar beet and 139 ha of wheat (point D) yields the highest margin and uses all the land. It is thus the optimum combination given the level of fixed resources available.

This simplified example illustrates all the important points of linear programming as a method of farm planning, and the method can be used to formulate optimum farming systems incorporating many enterprises as long as no additional fixed resources are needed. Occasionally, however, farmers contemplate radical changes which demand extra fixed resources. The motivation may be the need to re-quip an existing enterprise which offers the opportunity to reappraise the whole system or the stimulus may come

Table 5.4.

Point on production frontier	Enterprises		Total gross margin (£)
	Wheat (ha)	Sugar beet (ha)	
A	0	90	52,200
B	40	90	67,200
C	103	74	81,545
D	139	51	81,705
E	180	10	73,300
F	180	0	67,500

from outside, especially from changing government policy. For instance, the incentives offered by the EEC to reduce milk production (in order to reduce the Community's dairy surplus) recently combined with falling margins for milk to make many dairy producers consider alternative enterprises.

Changes requiring investment in new facilities are complicated by the need to estimate the likely gross margin not just for one or two years, but for the entire life of the acquired assets, which may be as much as twenty years. This exercise is imprecise even with the most sophisticated analytical techniques, since the farmer is effectively trying to forecast government policy and world events which may affect his costs and margins. Most governments have some short-term mechanism for guaranteeing farm prices, but there is rarely any long-term security in such measures. However, Table 5.5 shows that the relative profitability of various commodities has not changed significantly for many years in the UK, which reflects the remark-

Table 5.5. Index of gross margins for selected farm enterprises 1962–80 (winter wheat = 100) (Based on data in *Farm planning data*, Cambridge University, various years)

Enterprise	1962	1966	1969	1975	1979/80*
Potatoes	163	222	224	231	309
Sugar beet	180	219	208	232	186
Winter wheat	100	100	100	100	100
Spring barley	77	75	77	70	71
Oilseed rape	–	–	38	86	108
Field beans	38	56	76	54	78
Dairy cows (High yield)	142	146	215	235	162
2 years trad. beef	41	42	45	–	49
Sheep (lowland)	47	40	36	33	55

* The figures for 1979/80 are actual; all others were estimates issued in advance to guide farmers in determining farm enterprises. Such estimates are no longer issued.

able consistency of agricultural policy to which an earlier chapter referred.
Using these past margins as a guide for the future is a hazardous business, as farmers learned to their cost in the 1970s when rising oil and fertilizer prices made nonsense of previous experience. It is nevertheless the best guide they have, and all possible long-term investments have to be appraised in relation to projected gross margins in the full knowledge that, however thorough the calculations, there is inevitably risk involved in the decision.

Risk and uncertainty are inseparable from farm management, mainly because farming is such a long-term business, demanding investment which may only begin to produce returns many years later. In most industries it is possible to know with relative certainty the size of the market, the cost of inputs and, most important, the exact size of the output, so it is possible to calculate returns with comparative accuracy. In agriculture there is no such security. Even in relatively short-term enterprises like cereal growing (where a crop is sown and harvested in only a few months), it may be possible to determine very accurately the variable costs by buying seed, fertilizer and chemicals in advance, but the crop yield remains uncertain because it depends on the climate. Equally uncertain is the price at harvest, which depends on the size of the world harvest (which depends on the climate and even political upheavals) and on government policy. The returns are therefore very difficult to forecast, so the decision to grow wheat is made in relative uncertainty. Where long-term enterprises like livestock rearing are concerned, the uncertainty increases and the risk of losses grows. The decision nonetheless has to be made, but this entrepreneurial role is not one generally associated with farmers, though it is one which the circumstances of the industry oblige them to play.

Training and experience do help to reduce the risk inherent in farm decisions, but they cannot eliminate it. Suppose, for example, that a farmer with experience of growing spring barley decided to grow winter wheat instead – not apparently a drastic change. The published gross margins for the two crops might be:

	Average	Target
Winter wheat	£398	£540
Spring barley	£282	£448

If he achieved the target level for wheat, as he has done for barley, he would obviously increase his return. However, if his inexperience of wheat meant he achieved only the average gross margin – which is not unlikely, he would clearly be worse off for the change. The final choice of crop therefore depends on his willingness to take a calculated risk, since even the best farmer cannot know with certainty how a crop will perform on his farm until he tries it. Good management, in short, needs some good fortune to supplement it, and no training or experience can supply this.

THE PRODUCTION DECISION

The volume of a commodity produced is affected by the yield per unit of production (hectares or head of livestock). Yield is related to the amount of variable inputs allocated to the production process. The yield of wheat per hectare is related to the amount of nitrogen fertilizer applied; the yield of milk per cow to the amount and quality of food provided. The farmer may decide to aim for a high yield on a limited hectarage (or from a limited herd size), or a lower yield on an extensive area (or a larger herd). There is therefore a relationship between the determination of the area to be allocated to each enterprise and the yield which is aimed for, and this is affected by the relative prices of the fixed and variable factors. For the purposes of the following account, however, the amount of variable inputs to use will be examined in isolation from the use of the fixed resources discussed above.

What distinguishes the variable from the fixed resources is their availability. It is assumed that the variable inputs are readily available at a price, and that the farmer is able to purchase as much or as little as he requires. They can moreover be bought in small units, so the farmer can control to the nearest kilogram the amount of fertilizer applied to a crop or feedingstuff fed to a cow. This is true of many production inputs besides fertilizer and feedingstuffs – e.g. seed, agrochemicals, even casual labour or contractors' services. How much of each is necessary to achieve the desired output is the farmer's most delicate management decision, and it is complicated by the need to determine how to combine the variable inputs so that the chosen output is achieved at least cost. (Least cost is not always the principal objective. Low input of labour or artificial fertilizers and agrochemicals are equally important objectives for some farmers, and as in the calculation of the gross margin, whatever criterion is important is used to determine the optimum mix of inputs.)

These decisions are shaped by two interrelated sets of factors:

1. The technical relationships between inputs and outputs (e.g. the relation between increased nitrogen application and crop yield).
2. The economic relationship between the cost of inputs and the price received for the output.

The technical relationships of many production processes have been established by scientific research, and are readily available to farmers in textbooks, the farming press, and through their advisers. The farmer therefore applies these known factors fairly mechanically to his own farm, and his main management function lies in the less straightforward evaluation of the economic decisions. Here again he may be helped by economic theory, learned at college or university, or acquired from the same sources as the technical data, but the interpretation of the economic data is less easy.

The basis of his calculations is the production function, which is the relation between the amount of a single input and the yield of output, shown in Fig. 5.3. This shows clearly that there is a level of input at which output actually begins to fall, so there is a ceiling beyond which the addition of further inputs is unproductive, and may even be counterproductive. Nitrogen fertilizer applied to grass, for instance, scorches the grass above a certain level. There is thus a technical optimum at which output is at a maximum. However, the *economic* optimum (i.e. the level at which profit is maximized) is below this technical optimum. The farmer will therefore use

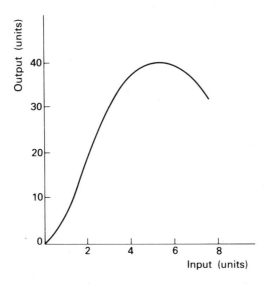

Fig. 5.3 The *production function*, showing the relationship between output and input

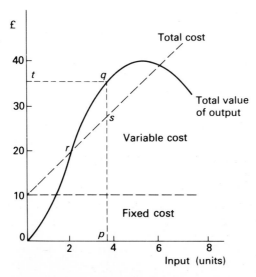

Fig. 5.4 Determination of the *economic optimum* (input level at which profit is maximized)

more fertilizer, say, as long as the value of an extra unit of output exceeds the cost of the extra unit of input, and he will stop when the two are identical – where the profit is in other words greatest. This economic optimum is shown at point q in Fig. 5.4, where the difference between the total value of the product (i.e. quantity × price) and the total cost is greatest. The farmer would then use p units of input, and his yield would be t.

Again, it is not sufficient to calculate the effect on yield of a single input, since the farmer needs to determine what combination of inputs to use in the production of a commodity. Technical factors are relevant insofar as the production of most commodities requires a minimum level of all variable inputs, but once this minimum is satisfied it is possible to substitute one input for another within certain limits. Since the farmer is generally looking for the least-cost combination of inputs, he considers their relative costs and uses as much of the cheapest before using those which cost more. Much of the time it is impossible for him to arrive at an optimum solution, since so many variables are involved. Computers have been used to improve the accuracy of the analysis, and have been successful, for example, in designing least-cost compound feedingstuffs for animals. Frequently, however, the underlying relations are not understood, so the most sophisticated techniques still give only approximate solutions.

It is now possible to see the effect of technical changes on farming systems. If, for instance, research establishes that the same yield of a crop can be obtained by using a lower-cost combination of inputs, the farmer has two choices. Either he can aim for the same yield at lower cost, or for a higher yield at the original cost. Both will increase his gross margin for the crop, so more land may be allocated to this crop, with consequent implications for the total farm system. Even without technical change, a fall in production costs will raise gross margins and lead to similar adjustments in the enterprise mix. Of course, the farm system may also be as much modified by an increase in revenue (and therefore gross margin) as by technical change or input costs. The farmer is therefore equally interested in ways of increasing his revenue from farm enterprises, and it is here that the marketing decision enters into his planning.

THE MARKETING DECISION

What to produce already involves, we suggested, a marketing decision, for it is pointless to produce a commodity for which no market exists, or one which brings only a poor return. Beyond this, however, the farmer has to decide how to sell what he produces, and this is not such a straightforward decision as it may initially appear. The decision includes consideration of the form in which produce is sold and the timing of the sale as well as to whom it is sold, and each of these decisions is related to the others. In all of them the farmer has a range of choices, and his ultimate choice is again based largely on the different costs and returns of each. These costs and returns may be expressible in terms of money, but they are often measured much less precisely – but no less importantly for all that – in terms of convenience.

For some specialized commodities the marketing decision is taken when the production decision is taken, often because production is linked to a contract to sell to a specified buyer or through a specified channel. Contract vegetable production for the frozen foods industry was mentioned in Ch. 4, and there are other farm products whose sale is similarly tied by contractual arrangements. All sugar beet in the UK has to be sold to the British Sugar Corporation, which not only controls how much may be grown, but when and where it must be delivered. There are also marketing boards for milk and wool which have no control over the volume of production, but which specify to whom the product is sold and (administering government policy) at what price. For most other commodities the farmer has a number of outlets to explore. Wheat eventually ends up in the hands of a flour miller or animal feed compounder, but farmers can sell direct to these firms or to a merchant or commission agent who re-sells to them. Similarly all livestock eventually end up in an abattoir, but they may reach it directly, through a direct sale from farmer to abattoir owner, or indirectly through the auction mart or a livestock dealer. The choice is partly dependent on the existence of established channels for different types of farm product. Breeding stock is generally sold through auction markets, for instance, while pigs sold for bacon curing are almost invariably sold direct on contract to the processors.

Other factors affecting the marketing decision are financial in character. What is the likely price in different markets? Are there costs involved in using one rather than another (e.g. commission paid to agents, market charges to auctioneers, extra transport costs, etc.)? Does one guarantee quicker payment than another? Farmers have to try to evaluate these factors in order to estimate their likely return from alternative markets, and a major consideration affecting their range of choice is often the location of their farms. Location close to urban centres has always offered some farmers market opportunities which others lack, one such opportunity which recently offered itself being the willingess of consumers to pick their own fruit and vegetables in return for lower prices. Farmers located near towns are also able to sell direct to consumers through roadside stalls and farm shops, and the popularity of these shops has led some farmers to change their enterprise mix in order to produce commodities suited to this market. More recently some livestock producers have opened freezer centres on farms to sell meat direct to consumers, and some have taken the exercise a stage further by opening shops in town centres to sell 'farm-fresh' produce. All this can increase the farmer's margin by eliminating intermediaries who previously took a share in the revenue as their payment for handling the sale.

The revenue from produce may be as much affected by the form in which it is sold as by the market in which it is sold. Grain may, for example, be sold directly from the combine harvester, or it may first be dried and cleaned, but the cost of drying and cleaning and storage would obviously have to be reflected in a higher price. Livestock may be sold *on the hoof* (alive) or *on the hook*. In the latter case the farmer is paid for the actual carcase weight after slaughter, and the price received is an accurate reflection of the saleable value of his product. Where animals are sold on the hoof, buyers estimate the killing-out weight in advance of slaughter, but the evidence is that their estimates are not very accurate, so the price is a less accurate measure of the actual carcase value. This can obviously work to the advantage of the farmer if buyers overestimate carcase weight from animals on the hoof, and the continuing resistance to deadweight selling suggests that it does so more often than it works to his disadvantage.

The timing of sale is obviously determined for some farm products by their highly perishable nature (e.g. flowers, fruit and vegetables), but many products can be stored for very considerable periods. Most maincrop potatoes are stored, as is most grain, the purpose being to match supply and demand more closely, and in the process obtain a higher price. At harvest there is clearly a much greater supply than the market can absorb, so someone has to store it until required. Whether it is the farmer or someone else who stores it (e.g. merchants or processors), the storage costs must be recovered in the form of a higher price. The decision whether or not to store on the farm is thus taken in relation to the estimated costs and returns. The former are fairly easy to determine, but any increase in price is difficult to assess. The task is easier where governments intervene to regulate farm prices, since the prices recommended or fixed by them do rise as the season advances to take account of storage costs and to ensure that there is a fairly steady release of produce onto the market. Where no government support exists, the farmer has difficulty in forecasting the benefits of holding produce on the farm, so he may therefore be less inclined to do the job, which has consequences further along the marketing channel.

For many people marketing is synonymous with advertising, but apart from small classified advertisements in local newspapers by farmers who sell direct to the public, most farmers never advertise as individuals. Farmers' cooperatives and other organizations to which producers contribute funds do advertise agricultural products (e.g. the Milk Marketing Boards), and here the object is not to benefit the individual, but to expand the market for all producers. It is therefore difficult for farmers to assess the advantage they derive from such campaigns, but it is fair to say that they are generally sceptical about their effectiveness, and contribute funds to them with reluctance.

IMPLEMENTATION AND CONTROL

Policy decisions affecting the choice of enterprises, the optimum use of production inputs, and the effective marketing of outputs represent only part of the farmer's management function. These decisions need implementing on a day-to-day basis, and this involves perseverance as well as technical competence since many enterprises demand literally the day-to-day presence of the farmer, 365 days a year. This management function is often under-estimated because it is not always obvious. The management content of re-moving hedges, erecting farm buildings or ploughing up moors is readily observed, but the control function of a farmer leaning on a gate watching a field of sheep or sugar beet is less easily appreciated. It is nonetheless im-portant for the farm's efficient operation, however, for early detection of plant and animal diseases is essential. The successful manager in any enter-prise is one who knows that something is going to happen before it happens, and takes action to avoid trouble. He is always at the right place at the right time, ready for any contingency. In agriculture this is truer than it is in many other industries because of the rapidity with which natural processes like disease and parasite infestation of crops and animals develop. Even a few minutes of rain at the wrong time can destroy a year's work, so the farm-er who does not take time to lean on a gate to keep his eye on things is inviting trouble.

Equally important are financial control and budgeting. The first is often delegated to a farm secretary – often the farmer's wife, who is responsible for paying bills and ensuring that payments for output are collected with the minimum delay. Budgetary control involves the comparison of actual finan-cial performance with the budget for any enterprise, and it is the most im-portant control measure in the long term, since it shows up poor physical and financial control. It is a measure of how well the farm system is per-forming, and the records kept in the process form the basis of future man-agement decisions. Continual examination of the farm's performance is it-self an integral part of farm management. The efficient farmer keeps records of technical and financial performance and compares them with information from other farms collected and published by the universities and advisory services. If he is satisfied with his own performance he may well make no changes to the farm system, but if it compares unfavourably with other farms, or if his own targets are not being met, he will need to examine his system critically in the search for improvements.

In this search, he will consult other farmers, study articles in the farming press, consult advisers and possibly scientists, and he must have sufficient data about his own farm operations to form the basis of a real comparative appraisal. This may simply reveal that a technical management detail has been overlooked, or it may indicate the need for a complete technical and financial analysis of a totally new farming system. In this case he will be able to call on professional assistance, but the final decision will still be his. The decision made, its implementation may call for extra resources or the ac-quisition of new farming skills, and certainly renewed attention to the everyday control which will reveal any weakness in the new system. For at the end of this, the process of evaluation of performance begins anew, and any new adjustments made to a system can never be regarded as definitive. On this realization, in the end, depends the success or the failure of the farm business.

6

Arable farming

Arable farming is the growing of all crops which require cultivation of the soil. The sequence of operations involved is familiar to every gardener: the soil is disturbed to produce a good growing medium; seeds or plants are set, and subsequently harvested. In principle it is a simple exercise, but as every gardener also knows, in practice it becomes much more complicated. For instance, there is no obvious reason why the same crop should not be grown year after year on the same ground. However, continuous cropping is subject to such complications that it is successful only if high standards of management are maintained. These complications arise mainly from biological constraints external to the crop, which the present chapter outlines with reference to cereal monoculture. The other biological principles which shape crop management are then discussed in relation to two other arable crops commonly grown in temperate regions – sugar beet and potatoes, since in both cases the principles of husbandry have been so well established that the farmer's art has been transformed into something like a science.

CULTIVATION AND HUSBANDRY

The range of crops available for arable systems is very wide, and represents no real constraint in temperate or tropical climates. The most important crops are those of which we eat the seed or the root. The first include cereals – wheat, barley, oats, rice, millet, sorghum, etc., and the roots include potatoes, sugar beet, cassava, etc. (The potato is not strictly a root, but that is how farmers consider and treat it. It is in fact a swollen underground stem.) In addition there are all the plants of which we eat the *buds* – e.g. brussels sprouts; the *flowers* – broccoli and cauliflower; the *stems* – sugar cane, asparagus; and of course the *fruits* – strawberries, blackcurrants, etc. Tree fruits like apples and pears are not considered as arable crops because they occupy the ground for such a long time. The other crop frequently treated as an arable crop is grass which is sown and then ploughed up (called a short-term *ley*) as part of an arable rotation.

Cultivation of the ground is generally necessary to ensure that the seed or plant roots are in intimate contact with it, in order to obtain moisture and nutrients. The amount of cultivation required depends on the kind of crop (some requiring only surface disturbance of the soil while others need a deeper cultivated layer), the soil type, expected yield and harvesting method. Under conditions of settled agriculture the normal practice is inversion by a plough followed by weathering of the exposed soil and further surface cultivation. Ploughing has the advantage of burying debris from the previous crop and any manure which is applied. It opens up the soil, allowing air to enter and making it easier for plant roots to penetrate and gain anchorage. The weathering and further cultivation break the soil into smaller particles, which ensures that seeds are closely surrounded by soil, and this contact is generally increased by a final rolling. All this should ensure a good crop, but it demands considerable manpower, machinery and fuel. It also has disadvantages. It may cause the soil to dry out, thus reducing seed germination, or it may compact the soil too much, preventing root penetration. Considerable skill and judgement are therefore necessary in timing and carrying out the operations.

To reduce these problems and the cost of cultivation, techniques have been developed of sowing cereals directly into uncultivated soil. This shows good results on well-drained loams and chalk and limestone soils, especially in dry years, and it can reduce the energy and labour requirements substantially. (Ploughing, drilling and cultivation demand some 3.9 manhours/ha compared with 0.6 manhours/ha for direct drilling.) However, if direct drilling is to work satisfactorily (and even then yields tend to be lower and more variable than with ploughing), debris from the preceding crop must first be removed, generally by burning. This is good farming practice, since it reduces diseases and weeds without using costly and potentially harmful agrochemicals, but it annoys neighbouring residents and has also been critized as being wasteful, since it is believed that the straw would otherwise be used for animal bedding and feeding. In fact, changes in livestock management techniques have very much reduced the need for bedding straw, and though it can be conditioned for use as an animal feed, its nutrient status remains low. The most serious problem associated with straw burning is therefore the potential danger to hedgerows and wildlife if it gets out of control. However, if chemicals are used instead to destroy crop debris, and the pests and diseases which overwinter on it, the natural environment may also be harmed, and the only advantage is that neighbouring humans are spared the smoke nuisance.

Once cultivation is complete, husbandry consists mainly in the control of pests, diseases and weeds which together reduce crop yield. The main groups of pests are the nematodes, insects (which cause direct damage or serve as vectors for virus diseases), molluscs and birds. The diseases are mainly viral or have fungal origins. The viruses are normally vectored to plants by biting insects, particularly aphids, whereas fungal diseases can be transmitted via the soil, the seed, by airborne means from already infected crops, or vectored by insects. Infected seed is in principle the most serious, as it could allow diseases to be transmitted over extensive areas, but there

are officially regulated seed inspection and certification schemes which virtually eliminate the risk. Control of the other problems remains the responsibility of the individual farmer, and his failure to take action can affect
others besides himself. If he fails to control pests and diseases in the soil the
damage will probably be confined to his own crops and income, but if he
fails to take adequate action against airborne diseases and flying insects, his
neighbours will suffer equally, and perhaps even more. What may seem a
simple personal choice not to spray his own crops because he dislikes chemicals may thus have repercussions in the form of responsibility for other farmers' losses, which make the morality of the choice much less clear.

   The control of pests and diseases is so complicated that the complexity of
the problem can only be suggested here. Some pests attack all crops indis-
criminately – for example, slugs and pigeons, but some pests and nearly all
diseases are species- and even variety-specific. The eelworm which attacks
potatoes does not attack sugar beet, and vice versa. Some fungal diseases exist in thousands of races, each specific to a particular variety of a crop – e.g.
yellow rust of wheat. Control is also complicated by the fact that although
diseases develop rapidly only in the preferred species, they can survive on
others and have intermediate hosts. The fungi and cyst-forming eelworms
also reproduce by spores or cysts which are highly resistant to natural de-
structive agents, surviving frost or drought for many years, for instance,
and becoming active again only when the host crop is present. Weed seeds
have also been shown to survive for a thousand years in archaeological sites
and still produce viable plants when uncovered. A weed is of course any
plant in the wrong place, so a barley plant in a wheat field is a weed, and in
the natural situation a whole range of weeds will invade crops. Once man in-
tervenes, by grazing or any form of cultivation or selective control, the natural flora balance is disturbed and certain plants flourish while others suffer.

   To control weeds, pests and diseases, agrochemical sprays are widely
used, but they are by no means the farmer's only weapon. Resistant
varieties of plants developed by plant breeders have proved a significant
benefit. Resistance to the four main leaf diseases of wheat alone has been
estimated to have saved the UK farming industry £170 million in the 1970s.
The other widely used biological control is crop rotation, whose value is less
easy to express in financial terms, though it is still the most effective way of
controlling many pests and diseases, and the *only* way of controlling some.
In any farming system, the same pattern of intervention year after year discriminates in favour of some weeds, pests and diseases. Gardeners know, for
example, that regular mowing of a lawn every year with the same machine
discriminates in favour of low-growing grasses and progressively eliminates
the taller-growing ones, which never have time to produce seed to reproduce themselves. Gardeners are also familiar with the build-up of disease in
the soil where certain crops are grown continuously, for example, potatoes
and brassicas. On a field scale the problem is intensified, and the farmer's
efforts to control the build-up of pests and diseases actually perpetuate the

problem, since once the competitors of one organism are eliminated, it is strengthened and competes more strongly with the crop, whose yield progressively declines.

CONTINUOUS CEREAL GROWING

The complexity of the problem can be illustrated by the commonest continuous cropping system of temperate regions – cereal monoculture. Efforts to grow cereals continuously are a high priority, for although man may not live by bread alone, the cereals could go a long way towards feeding the world's growing population. From an economic point of view they are an excellent arable crop, being relatively high-priced and readily saleable. There are winter and spring varieties, so if weather conditions prevent autumn sowing, a crop can still be planted in the spring when the weather improves.

Their husbandry is moreover simple. For winter crops the ground is ploughed 15–25 cm deep in late August, cultivated to produce a rough seed bed, and fertilizer is applied (40–90 kg/ha phosphate and potash, depending on the soil's nutrient status, supplemented by up to 20 kg/ha nitrogen if high yields are desired). The field is then drilled in rows with the seed, sometimes leaving *tramlines* (missing rows) along which tractors can subsequently drive without damaging growing plants. Finally the field is harrowed to cover the seed, and lightly rolled. In the spring the crop is rolled again to compact the soil around the plants to encourage the formation of *tillers* (grain-bearing side-shoots), and 75–140 kg/ha nitrogen are applied to stimulate growth. Thirty years ago no further work would have been done until harvest except for some weeding, but today the crop is frequently treated with chemical sprays to control weeds, pests and diseases. Together with improved cereal varieties, this treatment has increased yields dramatically – from 2.4 tonnes/ha in 1950 to 5.7 tonnes/ha (provisional figure) in 1980. In late summer the grain is harvested mechanically and may be dried before being stored in silos ready for sale. Straw remaining on the field is then baled for sale to livestock farmers or – more likely – ploughed in or burned, and the field is ready for the sequence to begin again.

The pattern for spring-sown cereals is substantially the same, though the entire operation is of course telescoped into a shorter period, and details like fertilizer application rates vary slightly. In neither case does the system require very skilled labour, but it does need machinery. As noted earlier, however, this can be hired from contractors, so no labour is needed except the farmer's, and no investment in machinery (nor even storage facilities, since grain can be sold immediately or stored in silos owned jointly by farmers). The system can thus be run almost single-handed by the farmer, using outside contractors at peak periods of cultivation and harvesting.

Since cereals are only modified grasses, and grasses can thrive for many years on the same land, there also seems to be no obvious biological constraint on the system. In fact, however, cereal monoculture encourages the build-up of pests, weeds and diseases, whose control is increasingly expensive of time and money. The most serious pest is cereal root eelworm, which reproduces itself by cysts capable of surviving many years in the soil. The eelworm invades and destroys the plant's roots, so its damaging effects are worst on light soils which dry out quickly in dry years, and on soils of low fertility where maximum root growth is needed to collect the necessary plant foods. Some control of eelworm is possible by chemical means, but on a field scale it is difficult to achieve even and effective penetration. The only completely effective control is not to grow the crop affected by the pest until all its cysts have died, but this obviously destroys the continuous cropping pattern. The alternative is to grow resistant varieties bred by plant breeders, but the eelworm quickly overcomes this resistance, so there is an ongoing battle between the plant breeder and the eelworm, and the farmer has to keep changing to new varieties and new treatments in order to maintain yields.

Fungal diseases attack all parts of the plant – roots, stem, leaves and ears. The most serious fungal diseases of cereals are *take-all* and *eyespot*, which are soilborne, and mildews and rusts, which are airborne. As its name indicates, take-all affects all cereals, though in some areas the subspecies which attacks oats is not present, so that crop may be grown in place of other cereals. Since in most developed countries the market for oats is very small, however, the revenue is lower than could be achieved by growing barley or wheat. The take-all fungus lives on live plants and decaying

Fig. 6.1 The effect of take-all on the roots of mature plants, showing (left to right) mild, intermediate and severe infection (Crown copyright reserved)

material, and on some grass weeds commonly associated with cereals. It attacks plant roots, and can reduce yield by as much as a third (Fig. 6.1). Control is achieved by using spring-sown cereals which do not occupy the ground as long as winter varieties. During the 6-month period that the soil lies bare it can therefore be cleared as thoroughly as possible of plant debris and weeds on which the spores overwinter. This is most effectively achieved by burning the straw on the field after harvest, followed by physical and chemical treatment of the soil. The alternative is to abandon cereals temporarily and introduce a break crop, so that the fungus is deprived of its preferred host. In this case, however, the break may not achieve such clear benefits as it does in the control of cereal root eelworm. Under a continuous cereal system take-all increases to a peak and subsequently declines to a lower level. This level is high enough to reduce the normal crop by a third or even a half, and in some years (especially wet ones) it can result in total crop failure. The break crop has the effect of reducing the take-all immediately, thus increasing the yield of the first cereal crop which follows the break, but thereafter the presence of the fungus increases dramatically and crop yield falls below the previous plateau level, which is regained only after several years. The total benefit of the break crop may thus be only a few kilograms per hectare, whereas a year's cereal production has been lost. The break crop therefore has to be very profitable, or have some other advantage to compensate for the year's lost revenue.

Eyespot, which is the other major soilborne fungus, cannot be controlled

Fig. 6.2 Farmers inspecting wheat varieties for their relative resistance to different diseases (Farmers Weekly)

by short breaks in the cereal cropping because it survives for many years on buried plant debris and straw. Its effects are devastating on all cereals except oats, so they may again be grown. Resistant varieties of wheat and barley have been bred, but their resistance is only partial, and the best control is to grow either early-sown winter cereals, in order to produce a strong plant with greater resistance, or late-sown varieties whose growth is faster. There are also prophylactic sprays which can reduce losses when they are used in combination with resistant varieties and appropriate planting schedules.

Airborne diseases have much wider effects than soilborne ones, and control again depends on a combination of good husbandry, resistant varieties and chemical control. Among the worst are yellow and black rust, which attack the leaves of cereals, reducing the plant area available for photosynthesis and thus reducing the grain yield. Yellow rust can reach epidemic proportions, as it did in the USA in the 1930s, when it decimated cereal production right across the grain belt. It exists in many races, and resistant varieties have had their resistance overcome within only a few years of introduction. Outbreaks spread frequently from *volunteer* plants, which grow from grain spilled at harvest which serves as a reservoir for the disease for the next crop. Strict hygiene in destroying all volunteer plants is thus vital, especially since there is no effective chemical control once the disease spreads to the growing crop. Chemical control of mildews is possible, but incomplete, so it needs supplementing with the highest possible standards of husbandry. In fact, the only diseases that can effectively be controlled by chemicals are the seedborne fungi, which can be destroyed by organomercurial compounds used to dress the seed. The method is cheap as well as effective, but these compounds are highly poisonous. Great care therefore has to be taken to cover the seed thoroughly when sowing, to prevent it being eaten by birds and small animals which would transmit the poison through the food chain to higher animals.

From this greatly simplified account it is already clear that what appears to be the simplest arable system – monoculture – is nothing of the kind. Further complications are introduced by weed competition, which robs the crop of light-energy and nutrients, and thus reduces yield. Chemical herbicides are now widely used to control weeds which used to be laboriously removed by hoeing, and cannot easily be removed mechanically. Since the 1940s growth regulator hormone weedkillers of the type regularly used by gardeners have been available to control the commonest annual weeds. These herbicides are specific, however, so a range of chemicals has to be used to control the different weeds which infest crops. The control of perennial weeds is more difficult because the most pernicious of them, blackgrass and wild oats, have the same plant structure as the cereals in which they grow, so it has proved difficult to develop sprays which will kill the weeds without killing the crop. These weeds are only now beginning to be treated successfully, and still at considerable expense. The worst grass weed of all – couch grass – can still only be contained by leaving the ground bare to permit a combination of physical and chemical treatment which weakens the plant. This means that winter cereals have not been widely used where couch grass is a serious problem, though the latest glyphosate herbicides now in common use offer better control.

MIXED CROPPING SYSTEMS

The complex problems associated with continuous cereals, or any other single crop system, are such that a majority of farmers in the UK still prefer to practise arable rotations which help to contain the build-up of pests, diseases and weeds. One of the earliest rotations was the Norfolk Four Course system which involved the use of clover and turnip break crops fed to sheep. If this rotation is used today, the livestock enterprise must be as profitable as possible, and until recently this invariably meant dairy cattle. However, dairying is a capital-intensive enterprise and requires skilled labour seven days a week, 365 days a year, and on an arable farm this can cause great difficulties as well as increasing labour costs substantially. Many farmers therefore look for a break crop which does not need a livestock enterprise, and one which also meets the following requirements:

(a) it should ideally be as profitable or more profitable than cereals, though a lower gross margin is acceptable if the alternative crop confers other advantages;
(b) it should have different pest, disease and weed spectra in order to reduce the problems of control;
(c) it should have labour peaks and troughs complementary to cereals, so that the same labour can be used for both;
(d) it should be capable of mechanical harvesting, preferably with the same or modified machinery;
(e) it should improve the land for the next crop.

The Norfolk Four Course answered all these requirements with its sequence of wheat, turnips, barley and clover, but the need for sheep in the system now makes it uneconomic. (See gross margins, Table 5.5.) On many soils a satisfactory substitute for turnips is sugar beet, and crops which can be substituted for the grass-clover mixture include other nitrogen-fixing legumes– peas and beans, or grass grown for seed production or for conservation (hay or dried grass). A whole series of rotations is thus possible, for example:

Wheat ... Sugar beet ... Potatoes ... Wheat ...
Wheat ... Swedes ... Barley ... Grass ... Wheat ...
Wheat ... Sugar beet ... Barley ... Potatoes ... Grass ... Wheat ...

The break crop *par excellence* in the 1970s was oilseed rape, sown in August or September and harvested the following June or July. It allows some spreading of labour since it is harvested before the main cereals and sown after they are harvested. In some years it may prove difficult to harvest the cereals and plough ready for the oilseed crop, and this delay can be critical because yield can be reduced by up to 60 per cent if it is not sown by the end of September. The machinery used for the two crops is similar enough to mean that relatively few extra fixed costs are incurred, and the pest and disease spectra are different. Oilseed rape, like cereals, nevertheless has specific diseases which build up as it features more frequently in a rotation. The main problems are caused by stem canker and downy mildew, but these can be controlled if only one oilseed crop is taken every four or five years, and if canker-resistant varieties are grown. When it is well grown, the gross margin of rape can equal that of winter wheat and is invariably better than spring cereals, so it meets most of the criteria for a break crop. Indeed it is such a good crop that it is increasingly regarded as an arable cash crop in its own right rather than as a strict break crop.

This has already happened in the case of sugar beet, which is the most profitable commonly grown crop in Britain and Europe, with the exception of potatoes. In Britain both crops are regulated by government quotas which limit the area on which they can be grown, the object being to restrict beet production to the capacity of the processing factories, and potato production to the amount necessary to meet consumer demand. Consequently the farmer is not free to decide how much of his land to allocate to them, nor how often they can be incorporated into his arable rotation. The other crops commonly grown as breaks or cash crops are vegetables on a field scale (carrots, onions, cauliflowers, peas, etc.), and some specialist crops like flower bulbs. Whether they can be introduced into the particular cropping scheme on any farm, and what position they occupy in the rotation, depends on many factors, some related to the physical environment and some to crop interactions. For example, cereals can be grown on quite shallow soils, but other crops which might figure in a rotation need deeper soil. Since most arable crops are mechanically harvested in the autumn, when it can be very wet, a light free-draining soil is also necessary for an all-arable rotation. The most intensive arable systems are therefore found on alluvial silts and loams, invariably in areas of lower rainfall like eastern England.

One of the most profitable short-term cropping patterns of recent years has been wheat, sugar beet and potatoes, and the technical factors affecting the yields of these crops have been exhaustively investigated. The yield of any crop is related to four interrelated biological factors within the farmer's control: sowing date, soil nutrient status, plant population, and disease, pest and weed control, the effects of which have been very thoroughly quantified for sugar beet and potatoes.

SUGAR BEET PRODUCTION

Sugar beet is a biennial plant which accumulates food in the root in its first year, ready to produce a flower and seed in its second year. For sugar production it is harvested in the autumn of the first year. The seed produced naturally by the plant produces not one strong plant, but several, so much laborious hoeing used to be necessary to thin the crop. This labour has been eliminated by producing a monogerm seed which is sown mechanically in April, and produces single plants which are also harvested mechanically. The method of mechanical lifting demands very careful preparation of the seed bed, the technique being to remove the top leaves before lifting the roots with a share onto a conveyor belt attached to a trailer. If this topping is to be done with minimum waste, the seed bed must be very flat, and there must be no stones to damage the lifting machinery. These factors determine where the crop can be best grown: the land must be flat and relatively stone-free, and artificial levelling of the natural contours is sometimes carried out to achieve this.

Research has shown that the highest yield of sugar is obtained from crops sown between 20 March and 10 April, but the actual date is determined by soil moisture content and by soil temperature, since sugar beet seed germinates slowly when the soil temperature is below 5°C (Fig. 6.3). High sugar yields are also dependent on high soil fertility, and on mineral soils the recommended fertilizer requirements are:

Phosphate	60 kg/ha
Potash	200 kg/ha + sodium 375 kg/ha
or	
Potash	125 kg/ha + sodium 750 kg/ha
Magnesium	100 kg/ha

On organic soils more phosphate and potash are required (100 kg/ha and 315 kg/ha respectively), but no sodium is necessary. On some soils where boron is deficient this has to be added. Potash and sodium substitute for

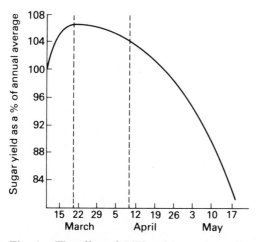

Fig. 6.3 The effect of drilling date on sugar yield, trial results 1963–75 (Direct communication, Brooms Barn Experimental Station)

88   *Arable farming*

each other within certain limits, and since sodium is cheaper than potash it is more economical to apply as much sodium and as little potash as possible. As Chapter 2 noted, however, an excess of sodium can be harmful, so the substitution needs to be done with care. The amount of nitrogen to be applied is complicated by the fact that it increases leaf area without increasing sugar content, and since the farmer is paid on sugar content and nitrogen is expensive, there is an economic optimum point above which it does not pay to apply nitrogen. Figure 6.4 shows that this economic optimum is at 120 kg/ha.

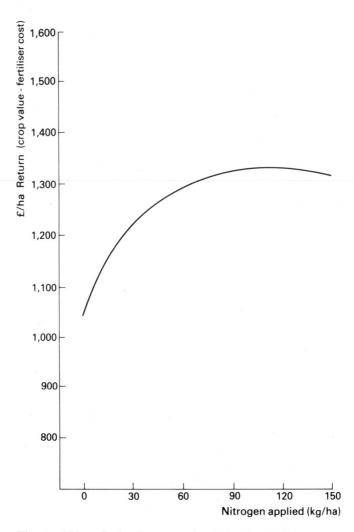

Fig. 6.4 Value of sugar beet crop after deduction of nitrogen cost at different application rates, results of trials 1978–80
(Direct communication, Brooms Barn Experimental Station)

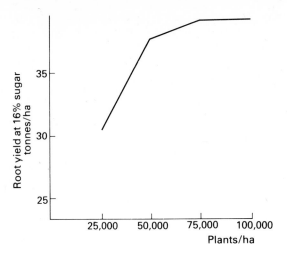

Fig. 6.5 Effect on sugar beet yield of a range of plant populations
(Brooms Barn Experimental Station, *Sugar beet: a grower's guide*, 1980)

The effect of crop density on yield is considerable, the maximum sugar beet yield being reached with a plant population of 75,000 per ha (Fig. 6.5). Plant population is determined by row width, seed spacing within the rows, and the percentage of seed which actually germinates. The germination rate of sugar beet seed varies widely from sample to sample, and if it is not indicated by the seed salesman the farmer is advised to have it tested in a laboratory. This is necessary because although an upper limit is placed on row width by the known fact that yield declines in row widths greater than 50 cm, below that the choice is completely open, except insofar as it is determined by the machinery used for cultivation and harvesting. Knowing the likely germination rate of the seed, however, the farmer can refer to tables provided by the sugar refiners which allow him to determine row and seed spacing precisely, and set his drill accordingly (Table 6.1). (Like the preceding figures, this table is reproduced from literature issued to growers on behalf of the British Sugar Corporation by an experimental station which specializes in sugar beet production.)

Similar detailed advice is provided on the control of weeds, pests and diseases. Chemical weed control in sugar beet is still difficult, but complicated procedures exist which must be followed very precisely. The two commonest pests are the most difficult to control – sugar beet eelworm and aphids. The former can only be controlled by rotation, and even a two-year break is insufficient to eradicate the organism, which is why the cropping scheme wheat/sugar beet/potatoes is not the most profitable in the long term. Aphids spread a virus which causes beet leaves to yellow, thus preventing photosynthesis, but this can be contained by good hygiene in combination with systemic insecticides.

Even with the best husbandry, the value of a good sugar beet crop can be

Table 6.1. Plant populations given by different seed spacings with different plant establishment percentages, based on 50 cm rows (Brooms Barn Experimental Station, *Sugar beet: a grower's guide*, 1980)

Plant establishment* (%)	14.2 cm	15.3 cm	16.5 cm	17.8 cm	19.0 cm	20.3 cm
			Plant population (1,000 s/ha)[†]			
90	120	113	108	100	95	87
80	106	100	95	89	83	77
70	93	87	82	78	75	73
60	80	76	71	66	63	59
50	67	62	59	56	52	48
40	53	50	47	44	42	40
30	40	38	36	34	32	30
20	27	25	23	22	21	20

* Plant establishment is defined as the number of plants, in early-mid June, as a percentage of the number of seeds sown; 70%, 80% and 90% establishment gives increasingly uniform plant spacing; 60% and lower establishment gives increasingly gappy and uneven stands.
[†] 1,000s plants/ha = number of plants per 20 m

seriously reduced by bad harvesting, which is not necessarily the farmer's fault. The harvesting of beet is coordinated by the sugar refiners, who issue delivery permits to farmers in order to maintain a steady inflow to the factories, which operate for only a few months every year (called the *campaign*, a name which indicates the military precision with which the exercise is coordinated). A farmer may miss his allocated delivery date because of bad weather which prevents lifting on time: only 30–35 good harvesting days on average have been recorded between September and the end of December, when the factories close. Lifting must be done quickly to prevent crop deterioration, but this can be further hampered by wet or frozen soil, so beet is often lifted and stored in anticipation of delivery dates. Even then the chances of losses are high, for the roots have to be stored in a frost-free condition.

POTATO PRODUCTION

Potato production is one of the most profitable large-scale arable enterprises in Britain. In 1979/80 the average gross margin for farms in a major potato-growing area was £1,231 per ha, compared with £738 for sugar beet and £398 for winter wheat. The fixed costs of potato growing are much higher, sometimes as high as £1,000 per ha, but many farmers still wish to grow them. However, the crop hectarage is restricted by the Potato Marketing Board (Ch. 12), and it is anyway not a crop that can be grown successfully everywhere.

Unlike sugar beet and many other crops, only 25 per cent of potatoes are processed (canning, freezing, dehydration, crisp manufacture) after they leave the farm, the vast bulk of the crop reaching housewives in its raw condition. The total household consumption of potatoes is almost static, so the growth in the sale of processed products has been achieved at the expense of the fresh product. On the other hand, the yield of potatoes fluctuates widely from year to year, mainly reflecting rainfall variation. (The 10-year average for the period 1970/71 – 1979/80 was 27.5 tonnes/ha, but the range was from 17.2 tonnes/ha in the drought year 1976 to 32.9 tonnes/ha in the wet year of 1979.) One consequence of this highly fluctuating supply and constant demand is that producer prices would fluctuate equally widely if left to market forces. However, the government tries to reduce price movements through the Potato Marketing Board by support-buying operations, as well as by regulating the crop hectarage.

Since the hectarage is restricted, farmers try to achieve the highest possible yield per hectare while simultaneously producing a product whose quality meets consumer requirements. The two objectives are by no means complementary, however, and this results in a series of compromises. From the consumer's point of view, a quality sample of potatoes is fairly uniform in size, easy to peel, does not blacken or disintegrate when cooked, and is not damaged, green, or distorted by secondary growths. Meeting these requirements depends on choosing an appropriate variety and treating the crop carefully during harvesting and storage, when most damage occurs. Yield is also related to variety, as well as length of growing season, plant population, nutrient and moisture levels, and disease and pest control. Production is concentrated on deep, highly fertile, well-drained soils which have a high moisture-holding capacity and relative freedom from stones. All the cultivation procedures are designed to produce a deep soil which can be ridged to cover the potatoes as they grow, thus preventing them from turning green (which indicates the presence of a toxin damaging to human health). It is also important to prevent the creation of soil clods which would limit the efficient penetration of soil-acting herbicides and cause damage to tubers and difficulties in harvesting. Stones in the soil are particularly damaging to tubers and machinery, so it is increasingly common for them to be removed, since this has been shown to speed up harvesting by as much as 50 per cent and significantly reduce crop damage.

There are three groups of potato varieties – early, second-early and main crop (though some can fulfil two roles). Early varieties can be planted in frost-free areas in late January/early February for harvesting in June, but the main crop is planted in April and harvested in September, after the cereal harvest and before the sugar beet. New and improved varieties are continually being produced, each with special characteristics. Some are well-suited to processing, high-starch/low-sugar varieties being preferred for crisp and chip making, while canners require smallness and uniformity of size, low starch and waxy flesh which withstands the processing treatment. Other varieties are chosen for their resistance to drought or disease. Farmers select a variety which suits their chosen market, but it must also be high-yielding, and where the two objectives conflict, there is no doubt that varieties are more often chosen for their high yield rather than their quality. In fairness to farmers it should nonetheless be said that government policy in the past encouraged them to produce quantity rather than quality, and that most consumers have always put cheapness before quality, resisting higher prices for better quality potatoes.

The longer the crop is growing, the higher the yield is likely to be. Earlies produce lower yields than main crops, but it is possible to lengthen the 'growing' season by allowing tubers to sprout before planting. This is essential for earlies, if reasonable yields are to be achieved in time for the highest market prices at the start of the new potato season, since prices fall sharply within even a week of the first deliveries reaching the shops. Even for main crop varieties sprouting before planting increases the crop by something like 3–5 tonnes/ha. Yield is also related to plant population, but the planting rate necessary to produce the highest economic return is this time difficult to determine, since it is affected by many separate factors, which include the size and variety of the seed potatoes. Most important, however, is the ratio of the price of seed to the expected price of the crop when it is harvested nine months later.

Potatoes are a very hungry crop, needing high soil fertility to produce good yields. Farmyard manure is applied wherever possible at a recommended rate of 40 tonnes/ha, supplemented by inorganic fertilizer, the suggested rates being 100 kg/ha nitrogen, 150 kg/ha phosphate and the same of potash (increased respectively to 150 kg/ha, 200 and 250 kg/ha where manure is not available). These are recommended rates, and there is a wide variation between farms in the rates actually applied. It is nevertheless clear that high yields are associated with high fertilizer levels. Since potatoes consist very largely of water, moreover, they respond well to irrigation. In some regions of low rainfall the yield has been increased by 1 tonne/ha for every centimeter of water applied, which has been enough to cover the costs of irrigation. More important, irrigation improves quality by preventing secondary growths. If too much water is applied, however, the proportion of dry matter may be reduced, so experiments have been conducted to establish quite accurately the crop's water requirement. The farming press publishes reports on soil water deficits in different areas, and from these farmers decide whether or not to irrigate. In dry seasons the ability to do so can be very valuable, since the overall reduction in the crop size will increase the price for those who have a crop to sell.

The scourge of potato production is potato root eelworm, whose effects can be clearly seen in Fig. 6.6. There are some partially resistant varieties, but the only effective control is a 4-year gap between crops. The two commonest virus diseases of potatoes – leaf roll and severe mosaic – are spread by aphids, and it is usual to spray crops to control the vector. The virus is also transmitted via infected seed potatoes, so the keeping of seed from one

Fig. 6.6 Effect of potato root eelworm on crop growth (Crown copyright reserved)

crop to another is discouraged. The production of uninfected potato seed is a specialist enterprise on farms in upland areas of the country which are too cold or exposed for the aphids to thrive, and the fields of growing potatoes

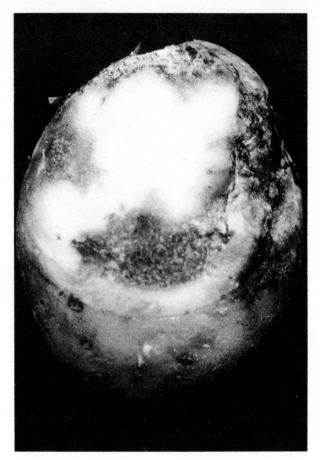

Fig. 6.7 A potato tuber affected by blight (Crown, copyright reserved)

are officially inspected for signs of virus disease. Only certified healthy seed
may be sold, and this constitutes a valuable cash crop for the seed raisers
and a reliable source of healthy tubers for lowland growers, an excellent example of the interdependence of farm systems.

The main disease of potatoes is blight, a fungus which is not carried on
the farm from year to year, so it has no rotational implications. It is
nonetheless a disease of major economic proportions, which was responsible
for the great potato famine which decimated the Irish population in the
1840s. The windborne fungus develops rapidly in wet or humid conditions,
and its progress can be plotted according to the weather. It generally appears in south-west Britain in June and spreads across the country at a predictable rate. Its arrival in an area can be forecast with very great accuracy,
following two consecutive days with minimum temperatures not less than
10 °C, and a relative humidity of at least 90 per cent for 11 hours or more
in each 24-hour period. Warnings are broadcast on radio farming program-

mes and in the farming press, and chemical sprays are used to try to limit its effects. The disease does not pass through the plant into the tubers, but if they are not completely covered with soil the spores fall onto the tubers, which rot later in the store. Most crops are therefore regularly sprayed with a fungicide, and to prevent diseased plant tops falling onto the tubers during harvesting they are destroyed two weeks before harvest begins.

Harvesting can be carried out mechanically, using a complete harvester which lifts the potatoes and deposits them in a trailer (Fig. 6.8). Alternatively the tubers are lifted mechanically onto the soil surface, then hand-picked by casual workers paid on a piece-work basis. It is increasingly difficult to find people prepared to do this laborious work, however, and some 70 per cent of the crop is harvested mechanically. This has the advantage that fewer tubers are left in the soil to act as a reservoir of disease for the following crop, but the machinery has to be well-designed to avoid damage to the tubers, since at this stage their skins are very soft and the flesh susceptible to damage. Much effort has been put into the design of efficient machinery, however, without which the potato harvest would become extremely difficult, given the shortage of labour to do the job.

SUMMARY

Arable crop production is becoming increasingly scientific. Highly complex specifications (*blueprints*) exist for the major crops, which farmers are exhorted to follow in the search for ever-higher yields. These blueprints are based on scientific investigation of the physiological processes of plants, the effect of pests, diseases and weeds on crop yield, and their control by cultural and chemical means. Machinery substituted for labour is constantly improved to work faster, using sophisticated attachments which improve the

Fig. 6.8 Potato-lifting machinery (Farmers Weekly)

quality of work and reduce damage to produce. All these developments increase the cost of production, but to be fair the cost would probably increase anyway because the cost of agricultural inputs like land and labour constantly rises.

The question increasingly asked by farmers and their advisers is not just whether these high-cost/high-output systems can continue to be justified in a context of dwindling resources, but whether they are even more profitable than lower-cost/lower output systems. Trials have been carried out on cereals and potato blueprints which indicate that the highly sophisticated systems are *not* the most profitable under ordinary farm conditions. Table 6.2 summarizes the results of one such trial which tested four winter wheat systems. Two systems (Laloux and Schleswig Holstein) are widely used on the Continent, and use high inputs and intensive growing techniques (like high plant population). The third system used conventional growing techniques with high nitrogen, while the fourth combined low inputs with conventional growing techniques, and the table makes it clear that this low-input system averaged a higher gross margin over the period of the trial than the Continental systems, and almost as high a margin as the high-nitrogen system.

Another report based on a longstanding experiment which tested under farm conditions the viability of the potato blueprint devised by a horticultural station has also concluded that the blueprint is not a financially viable alternative to standard techniques where irrigation is not possible and high management standards cannot be maintained. This does not devalue the work of experimental farms and stations, which have undeniably been in the forefront of efforts to increase output per man and per hectare, which must both remain high priorities where land and labour are at a premium. It does underline the need to recognize, however, that although sophisticated blueprints may produce the highest yields in experimental situations, on real farms they may prove disappointing, and farm margins fall. Research is therefore turning towards low-input arable systems which some farmers already claim are repaying their efforts better, and will probably repay them even more as the cost of inputs continues to rise (Ch. 10).

Table 6.2. A comparison of winter wheat systems in terms of their gross margin.
(Based on data from Agricultural Development and Advisory Service trials, 1978–80)

Growing systems	Average yield tonnes/ ha	Average revenue (£)	Average variable cost (£)	Average gross margin (£)
Laloux	8.11	736	171	565
Schleswig Holstein	8.25	758	224	534
High nitrogen	8.14	747	160	587
Low input	7.66	706	120	586

SPECIAL TOPIC: MECHANIZATION

All arable crops involve the use of machinery whose use has consequences for the farm system, and for the natural environment. Much of the machinery is non-specific – tractors, ploughs, etc., and some is common to a range of related crops. A combine harvester can be used for all the cereals, oilseed rape, grass seed, and even peas and beans. Other machinery is highly specialized, like potato ridgers or harvesters. The machinery used on a farm may have been acquired to make a chosen system possible, but often the system is shaped by the available machinery, since the cost of new machinery is high. The introduction of potatoes or sugar beet into a continuous cereal system, for example, entails the acquisition of special machinery requiring a high capital outlay, whereas if peas, beans or oilseed are introduced as break crops, relatively inexpensive attachments can convert existing machinery to handle the new crops.

The principal objectives sought by mechanization are:
1. Reduction in labour requirement.
2. Elimination of manual work.
3. Timeliness and improved quality of husbandry.
4. Higher output and better quality of produce.
5. Increased profit.

These objectives clearly interact, and though profit maximization might be expected to be the primary one, the others should not just be seen as ways of achieving this goal. A reduction in the labour used does not necessarily mean a reduction in overall cost. The elimination of manual work is often desirable for its own sake, whatever its effect on profitability. One of the best reasons for mechanization is improved timeliness in carrying out farm operations, especially where the labour requirements of different enterprises overlap, since this may increase yield or quality, or both. It may also allow an enterprise mix which would otherwise be impossible, and this may allow a change in the farm system which directly increases the overall income.

The most important interaction is between machinery and labour. All farms have labour peaks: on arable farms, at spring sowing and autumn harvesting. It is possible to spread these labour peaks slightly by a well-balanced crop combination, but some extra labour is still necessary for sowing and harvesting. Since casual labour is not generally available, farmers now have three choices: to change the system radically, which may reduce farm income; to employ more permanent labour or bring in contractors; or to bring in more machinery. The use of contractors is resisted by many farmers because their dependence on outside services can disrupt the farm schedule, and the shortage and high cost of permanent labour generally prevents the employment of extra workers. The use of more machinery may therefore be the only way of retaining a successful farming system. But every machine needs an operator, so the use of more machines may entail the employment of more labour. The answer is thus increasingly not *more* machines, but *bigger* machines which achieve higher work rates with the same amount of labour.

Recent statistics show that this solution is becoming increasingly common. The number of tractors in the EEC increased by only 42 per cent between 1965 and 1977, but the horsepower per 100 ha increased by 220 per cent. In the UK the number of tractors rose by only 2 per cent in the same period, but horsepower increased by 60 per cent. This use of increasingly large machinery has consequences for the farm, for the farming industry as a whole, and for the landscape. Machines with higher work rates (say, a sprayer with 15-metre as opposed to 6-metre booms, or a 6-furrow rather than a 3-furrow plough) are invariably larger and heavier, and need larger and heavier tractors to pull them. These machines damage the structure of certain soils by causing the compaction of the subsoil. This impedes drainage and requires still more powerful machinery to break up the underlying hard pan. Damage to soil structure may cause progressive reduction of crop yields, and some crops may have to be completely abandoned. The introduction of heavier machinery may also prevent, rather than improve, the timeliness of farm operations, since it cannot be used on wet land in case it damages soil structure. Problems of this kind ensure that machinery manufacturers are constantly improving the design of their products, developing better weight distribution or new tyres or tracks which spread the weight and thus reduce ground pressure. Farmers also innovate, modifying machinery to suit their individual circumstances.

The principal advantage of many mechanized operations is undoubtedly speed. The 3-stage sugar beet harvesters which handle 4 ha/day where single-row machines manage only 1.5 ha/day clearly reduce the time spent on a job which we saw has to be completed as speedily as possible to meet delivery schedules. They may also allow a winter crop of wheat to be sown if the ground can be cleared by the end of October, which brings a higher return than a spring cereal. These machines also generally perform jobs better than men performed them in the past, leaving less diseased debris behind, for example, to infect the next crop. Above all, however, they replace the men who refuse to do the tedious and laborious jobs they perform, and even with the increased cost of energy they have helped to restrain farm production costs where, as in the west, wages have increased faster than energy costs.

The problem is that as machines become increasingly complicated and more expensive to buy and operate, they demand more *skilled* labour, which costs more than unskilled labour. A farmer is unlikely to put an unskilled man in the cab of a £30,000 combine harvester, so the saving in labour costs may be reduced, but he is at least certain of getting the job done. However, if advantage is to be taken of the high output of large machinery, the system into which it is fitted must be highly organized. It is wasteful for a £30,000 combine to stand idle because there are insufficient tractors and trailers to cart grain away as fast as it is produced. An increase in the potential

Fig. 6.9 The alternative to mechanization: farm workers lifting swedes by hand (Farmers Weekly)

efficiency of one component of the system thus demands an all-round increase in efficiency, and it has to be said that on many farms the quality of management is not high enough to realize to the full the benefits of sophisticated machinery. Many farms in many developed countries are also becoming over-capitalized, and the only way to exploit this excessive investment in machinery would be to increase the size of farm holdings (unless farmers cooperate in a machinery syndicate which spreads the cost and use of machines). The pressure for larger farms therefore grows, and the entire structure of the industry may in time be modified.

There are two other ways in which mechanization affects the industry. In the first place, farmers with smaller crop areas do not purchase new machinery, but rely on second-hand equipment used for one or two seasons on a larger farm in good working conditions. However, the large machinery which is coming onto the second-hand market is too big for the small farms, and the question is where they will in future purchase their equipment. The other problem is that even fairly large farms cannot justify the purchase of the really large machinery now being used, and this is leading to the development of machinery cooperatives which jointly own equipment, a solution which could also help smaller farmers to purchase machinery appropriate to their holdings. The other alternative is the increasing use of contractors, which has the advantage that expensive machinery is in the hands of specialists, while farmers are released to do the jobs for which outside help cannot be found.

The impact of mechanization on the landscape is the main concern of the non-farming population. Large machines need large areas of land in which to work, so obstacles like trees, hedges, boulders, and even hillocks may disappear to make room for them. The arable areas of Britain are increasingly laid out in large, featureless rectangular blocks separated only by drainage ditches, and only then if it is too expensive to lay underground pipes. It is highly unlikely that the entire landscape will be transformed into empty wilderness, however, because arable farming is limited to a small area of the country, and the livestock and mixed farms elsewhere neither require nor are likely to change the landscape radically. There may be some amalgamation of small fields, but the real impact of mechanization has already been felt, and any further changes are certain to be opposed by the conservation lobby. The conservation case will not be helped by automatic opposition to any and every technical innovation, however. The landscape we cherish is in fact a farmscape, created by agriculture over many generations, and it is neither possible nor desirable to preserve it like some static monument. The creation of a landscape is a dynamic process, and farmers are arguably more aware of this than occasional visitors to the countryside, and not without cause do the majority of them call themselves curators of the countryside, who may be trusted with its care more than their critics think.

7

Grassland systems: introduction

Contrary to general opinion, the predominantly grass landscape of Britain is not the natural one. The natural climax vegetation is woodland, which was progressively cleared for agriculture over the centuries, creating the landscape which is now considered natural. Grass is not in fact something which grows naturally in fields, but a crop. It is indeed the most important crop grown in Britain, accounting for some two thirds of the total farmed area.

On many farms grass is the only crop which is grown. On a small minority of farms it is a cash crop grown for seed, for sale as hay, or for drying (for sale to feed compounders). In most cases, however, grass is grown for consumption on the farm of origin by livestock kept for meat and milk production. Even the cash crop finds its way eventually into animals, and grass as an animal feed is regaining its traditional importance as alternative feeds like grain increase in price. The evidence is nevertheless that its utilization is not as good as it could be. On most farms only about half the production possible with good management is being achieved, so the productivity of grassland systems is the object of intensive research.

Grass must provide animal food throughout the year, since the ruminants which convert it into protein need constant supplies of long fibre material to maintain their digestive function. Since grass growth varies with the season (being highest in spring and virtually ceasing in winter – Fig. 7.1) conserva-

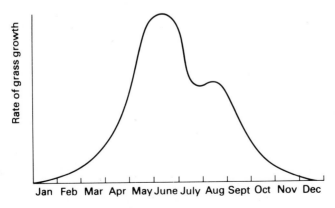

Fig. 7.1 Annual pattern of grass growth in the UK

tion is necessary in some form, generally hay or silage. These are fed to animals in the winter, and in most cases they graze in the fields during the rest of the year. In some systems, however, fresh grass is cut and brought to them throughout the year, and they are not allowed to graze the land directly. The reasons for choosing this or any other grazing and conservation system relate to the amount of nutrients which each makes available, the labour organization of the farm, and the farm system as a whole.

There are basically two grassland systems: hill grazing and lowland grass. In many upland areas grass is the only crop which the climate and topography permit, so it is absolutely essential to the continuation of farming in these regions. In lowland conditions grass is a crop choice open to any farmer. In an arable system it may be introduced as a break crop for the purposes of pest or disease control, in which case it is generally a temporary ley sown from seed for a maximum of three years. On other lowland farms grass may be the only crop for the simple reason that it is the most profitable. One of the most profitable uses for grass in the last thirty years has been milk production, and on many small lowland farms dairying is the entire farm system because it produces the highest income. Here, as in the uplands, grass production is an economic necessity. On such farms the pasture may be long-term leys which are ploughed and resown roughly every five years, but half the area of enclosed agricultural land is now composed of permanent pasture which is rarely ploughed and reseeded, and is instead improved by other methods.

GRASS PRODUCTION

Although we use the term 'grassland', pastures may be pure grass or a grass and clover mixture. Weeds are also present, though farmers use herbicides to exclude invasive or poisonous ones. Grassland is thus an ecological system. The grasses used are not generally the naturally occurring ones, which have lower yields than their cultivars, which were selected and bred to answer specific needs. Most were selected for their high blade-to-stem ratio, which increases their feeding value, but some have been developed for their earliness, because they grow stiffly (and thus cut well for hay or silage), or because they are prostrate (and therefore suit sheep, which cannot graze long grass). Even different strains within the same variety exist which are good for hay making, or alternatively, for grazing. The commonest grasses used in the UK are different varieties of perennial ryegrass, and Italian ryegrass, which is treated as an annual. Both establish quickly, producing good yields in spring and early summer, and respond well to fertilizer. Other commonly used grasses are cocksfoot, which is deep-rooting and therefore drought-resistant, and timothy, which has high palatability and is winter hardy. The main clover used in Britain is the white one, because it is longer-lasting. Red clover can also cause severe digestive problems in cattle, causing *bloat*, which can be fatal. Clover is vital for its high nutrient value but it is not as productive as grass, so it must not be allowed to dominate the sward.

Most pastures are a grass-clover mixture which may be carefully chosen to suit a given farm or particular grazing or conservation requirements. Some mixtures were formerly very complicated, containing up to ten varieties, but the trend is towards simpler mixtures of one or two grasses and clover. Recent research also questions the conventional wisdom that cultivars are more nutritious than naturally occurring grasses, which may for instance be richer in trace elements. The important considerations now are therefore to produce a sward which answers particular requirements (for intensive grazing, for earliness, etc.) on particular soils at an acceptable cost, and which will last as long as possible without losing its nutrient value.

Fertilizer application is important for grass production, especially when the grass is not grazed *in situ*, but cut for feeding to housed livestock or for conservation. The key to grass productivity is nitrogen, some of which is derived from the manure of animals grazing on the fields. The clover in the sward fixes more nitrogen in the soil, and the farmer supplies the rest in the form of inorganic fertilizer. There is an optimum relationship between the last two nitrogen sources, since heavy applications of inorganic fertilizer (in excess of 150 kg/ha) depress clover growth. The question is therefore whether to rely entirely on one or the other, or what combination to aim for. The decision depends mainly on the system in which the grass is integrated. In high-output dairy systems clover may be ignored, and high nitrogen dressings applied (up to 450 kg/ha). In low-output systems this may be unjustifiable, and clover will be included in the seeds mixture and lower nitrogen dressings applied. (A vigorous clover-rich sward can fix up to 200 kg/ha of nitrogen annually, given optimum growing conditions.) The entire output of the New Zealand sheep and dairy industries is built on clover-grass swards and low nitrogen input. In the UK a higher nitrogen input has generally been considered necessary because the ability of clover to fix nitrogen is related to soil temperature, and in Britain this is rarely high enough to produce sufficient nitrogen in early spring when it is most needed (because winter feed stocks are low or exhausted). It is therefore common to apply 75 kg/ha of nitrogen in spring to provide *early bite*, and a similar dressing may be applied in May or June, which is not enough to depress clover growth significantly. On many lowland farms the problem is solved by sowing some fields with a grass-clover mixture to provide grass for conservation and mid-season grazing, while other fields are sown with pure grass swards which can be fertilized to provide early bite. This is especially important where ewes are lambing and cows coming into full milk production after spring calving. These early swards are often a mixture of Italian ryegrass and forage rye, or pure Italian ryegrass, which grows at lower temperatures than most other grasses. After intensive grazing they are then ploughed up, and the field resown with a more permanent mixture.

In addition to nitrogen grass needs phosphates and potash. A non-deficient soil requires approximately 50 kg/ha of phosphates per annum, and in practice, since phosphates are not leached out of the soil, they are generally applied every third year. Potash is returned to the soil in animal manure, but it is still recommended that extra potash be applied at the rate of 50 kg/ha. When grass is removed in conserved products like hay, potash is naturally removed with it, so a dressing of about 25 kg/ha tonne of dry matter removed is necessary. The application of potash must, however, be decided in relation to two important factors. First, every 3 kg of nitrogen must be balanced by 2 kg of potash to support plant growth. There is also a relationship between potash and magnesium levels in the soil, and if potash is applied to grass in the spring there is a risk of producing the metabolic disorder *hypomagnesaemia* in grazing cattle.

Grass needs adequate soil moisture, and though rainfall in the UK might be thought adequate, it is actually insufficient to sustain growth in south-east England in eight out of every ten years. At such times irrigation will increase yield, which is seriously depressed when the soil moisture deficit exceeds 50 mm. On the other hand, although grass will tolerate badly drained soils, it will not flourish, so drainage should be improved wherever possible to increase the amount of available grass.

GRAZING SYSTEMS

Most grassland is grazed at some time of the year, and the farmer has to use good management to ensure that its productive potential is maximized. This depends on three main factors: the quantity and quality of the herbage, the number and potential productivity of animals, and the efficiency with which pasture is utilized. Throughout, the needs of animals *and* grassland must be kept in mind, for they are not always compatible. Especially important is the need to avoid permanent damage to the sward by poaching, which results from over-grazing or from grazing in very wet conditions.

The object of grassland management is to find a satisfactory balance of high productivity per animal and high productivity per hectare, which is not easy since the two are in conflict, as Fig. 7.2 demonstrates. This shows that as the number of animals per hectare increases (*stocking rate*), the yield of each individual falls, but the yield per hectare rises. The task is thus to find the optimum point at which the difference between revenue and costs is at a maximum, which determines the intensity of the grazing system.

Under *extensive* conditions a fixed number of stock is carried on a given area, with very little variation within the year or from year to year. No attempt is made to match stock numbers precisely to the amount of grazing available. This is the ranch system, which in European conditions is called *set stocking*. It is most successful with spring-calving dairy herds (Ch. 8), since their demand for grass is very similar to the grass growth curve. In the case of sheep, farmers control the lambing date so that lambs are born a few weeks before maximum grass growth is expected. On mountain farms lamb-

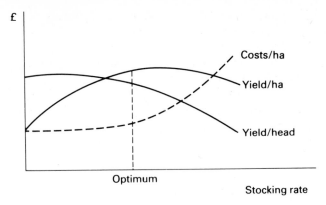

Fig. 7.2 Determination of optimum stocking rate

ing may be as late as May; on lower hills it may be the end of March. On low-land farms it can obviously be much earlier (Ch. 9 and 10). The synchro-nization of lambing and grass growth is thus achieved by altering the date on which rams are run with the ewes. The same principle holds for suckler calf production, but the freedom to choose a calving date is limited by the need to produce a 250 kg calf by the autumn sales. In practice farmers are therefore obliged to calve cows earlier and feed concentrates until grass growth accelerates. Under extensive conditions, in other words, the con-straint posed by the grass growth curve is compensated for by controlling the other components in the system.

Extensive systems are most frequently found in hill areas, and for sheep enterprises on most farms and traditional beef fattening. These are all low-output systems (see gross margins, Table 5.5), so they will not cover the high costs associated with intensive grazing systems. Some intensive sheep and beef systems exist, but they are not widely practised. The main live-stock enterprise in which they are practised is dairying. A dairy cow needs approximately 12 kg of dry matter daily. To provide this, 0.2 ha of grazing land/cow/year is necessary, given nitrogen at the rate of 370 kg/ha to supply sufficient grass for grazing, supplemented by another 0.2 ha/cow to provide grass for conservation. (Farms with insufficient land to provide this area for growing *winter keep* (hay or silage) are dependent on purchased supplies, or must carry less stock.) To meet these requirements three *intensive* grazing systems exist: paddock, strip and zero grazing, and 75 per cent of the dairy herds in Britain are managed under one of them.

Paddock grazing is the simplest intensive system, involving a large num-ber of small fields round which stock are systematically moved so that there is always fresh grass available in the growing season. This system ideally suits the small-field pattern of large areas of Britain. Where existing fields are too big, semi-permanent electric fences are used to divide them into smaller paddocks, each of roughly identical size, and each with convenient access and water supply (since each dairy cow needs 50 litres of water daily).

One paddock is generally grazed every day, and the appropriate stocking rate is 100 cows to every 0.8/1.0 ha. The length of the cycle before each paddock is grazed again is approximately 30 days, depending on weather conditions and the stage of grass growth. Where there is surplus grass some paddocks are taken out of the rotation and cut for silage. After grazing each paddock is dressed with about 50 kg/ha of nitrogen.

Strip grazing is common on small farms, but it is labour-intensive. An electric fence is used to provide a strip of grass sufficient to feed the herd for 12 hours, and then moved along to open a new strip for grazing. The width of the strip is related to the fact that a sward 20 cm high provides sufficient grass for 175 cows/ha/day. This system is infinitely variable. If the field is large, a second fence can be erected behind the cows to prevent access to the grazed area, thus allowing regrowth and preventing damage to the pasture. It creates a moving paddock, and is an extremely efficient way of using grass. Since there are no semi-permanent fences, the conservation of surplus grass is also easy, and the same field can be used for both grazing and conservation.

Zero grazing is the most intensive system, which brings the grass to the cows. This system is commonly practised on the Continent, either where climatic conditions increase the risk of poaching, or because farm holdings are split up over a large area, making it impossible to move cattle easily from field to field. In Britain the system has only recently been adopted, following the introduction of special machinery for cutting, carting and unloading grass direct to housed cattle without manual labour — indeed, with-

Fig. 7.3 Grazing cows restrained by an electric fence (Farmers Weekly)

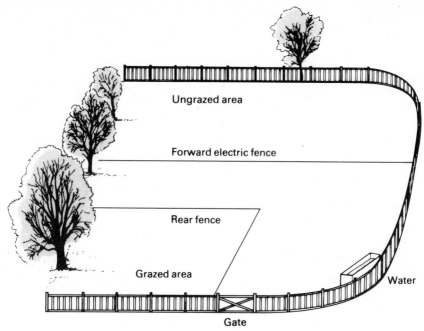

Fig. 7.4 Strip grazing

out the tractor driver leaving his seat. The system achieves very high levels of utilization, since no grass is spoiled by the animals' own droppings, and thus wasted when the field is next grazed. It also allows distant fields to be used for grass production, or fields without fences or water, and this improves the rotation on many farms. However, it is a high-cost system in terms of labour and machinery, and the additional cost has to be weighed against the potential gain of 15 per cent of grass fit for grazing which has been recorded under experimental conditions. The increasing cost of land and fertilizer may nevertheless increase the system's attraction, and this could accelerate the trend towards larger fields in grass-growing as well as arable regions.

All grazing systems have to be designed to minimize the presence of parasitic worms which infest animals and reduce their performance. Along with grass grazing animals ingest large numbers of larvae which grow from eggs deposited in the faeces of other animals. These parasites infest animals' lungs, liver, stomach and intestines. Stomach and intestinal worms alone include some thirty species of nematode, some specific to cattle or sheep, while others infest both. The life cycle of these worms is sometimes simple, but in some species it is complicated by the need for an intermediate host. Liver fluke, for instance, which is one of the commonest parasites of sheep, has a snail for its intermediate host, and the only effective way of containing the parasite is to eliminate, by draining or fencing off, the wet areas in which the host snail thrives.

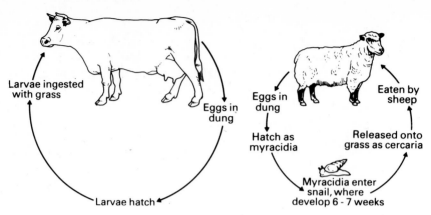

Fig. 7.5 (a) Life-cycle of lungworm causing *husk* in cattle
(b) Life-cycle of liver fluke via intermediate host

Parasites are always present in the pasture, and most animals have a continuous moderate level of parasitization without showing any symptoms, but in some circumstances the level of infestation flares up suddenly and severe losses may be suffered. Most parasites thrive in wet conditions, but some are triggered by very specific circumstances. One species of nematode needs a period of cold before it hatches, so it always gives a spring flush of infective larvae. The greater the farmer's understanding of such circumstances, the greater the likelihood of controlling parasites effectively, and control is essential, for although animals develop some resistance in time, young animals are highly susceptible. Effective control depends on good management of pasture and animals. The first priority is to prevent the initial contamination of clean grass. The second line of defense is to attack the parasite in its free state on the pasture, and to eliminate the conditions in which it (or its intermediate host) thrives. Finally, most animals are routinely dosed to contain any reaction once parasites have been taken up.

Intensive grazing can have the same effect on the build-up of parasites as continuous cropping on pests, and one remedy is again rotation. However, earlier claims that a rotation cycle of 28–40 days free of stock sufficed have not been substantiated, and it appears that clean grass which has not been grazed by the same class of stock for at least one year is essential if parasites are to be controlled. This is almost impossible for the ordinary farmer to achieve, so the animals are generally routinely dosed against parasites before turnout. In addition there are certain husbandry practices which help. Young stock, for instance, should never be turned onto land previously grazed by mature animals. If suckling calves and lambs are turned out with their mothers onto clean pasture, however, their resistance increases along with the rate of infestation, since, being mainly milk-fed, they do not ingest many larvae directly from the grass. If rotation is combined with the mowing of any surplus grass left after grazing the problem is further reduced, since parasites do not thrive in short grass.

CONSERVATION OF WINTER FEED

Most ruminants are fed on grass all the year round. Sheep can graze pastures for most of the winter as well as during the summer, but in most regions cattle are kept inside for at least part of the winter because of the damage they do to wet pasture. Some kind of conserved grass must therefore be provided to carry them through the winter, and in severe weather sheep are also fed hay to supplement the reduced grass growth.

In conserving grass for winter feed the aim is to provide a product whose feeding value is as high as the original herbage, which the animals will find palatable, and which has been produced with minimum waste. This can be achieved either by drying (naturally on the field, as hay, or in an industrial process which produces grass pellets), or by chemical preservation in the form of silage. By far the commonest is hay making, but much of the hay made is of indifferent quality. Its feeding value depends on the quality of the sward, especially its clover content; the stage of growth when grass is cut; and the weather conditions during the drying period. To make a high-quality product grass must be cut at the early flowering stage, when pollen is being shed, which occurs in early June in the UK. Farmers habitually delay cutting, however, so that a larger volume is conserved. This extra volume is nevertheless achieved at the cost of quality, since stems become woody as they stiffen and the leaf blades become less digestible. The feeding value of the hay is therefore reduced.

Hay quality also depends on the speed and effectiveness of the drying process. Since a standing crop dries faster than cut grass (because air still circulates through it), the cut should not be made until any dew on the grass has dried, and the mown grass should subsequently be turned as much as possible to circulate air through it. Thorough drying is necessary to reduce the moisture content below 18 per cent, or fungi develop during storage which not only reduce the hay's quality, but also release spores which cause a serious bronchial condition (*Farmer's lung*) when inhaled by humans. The annual yield of hay may range from 3 to 7 tonnes/ha, depending on the quality of the sward and its management. If it is cut early, it may be possible to take a second cut, or certainly to obtain some *aftermath* grazing.

Hay is normally baled mechanically to produce light, easily handled bundles which can be tidily stacked and readily carried to animals in the fields or in winter housing. The major obstacle to good hay making in Britain is the poor weather. A good crop requires 2–3 days of dry weather, but records show that between May and August there are on average only two such spells each year. Drying is therefore difficult, and the quality of the product frequently inferior. Even so, some 70 per cent of conserved grass is still produced as hay, although silage making is less susceptible to bad weather and also integrates better with some of the more intensive grazing systems. The problem is that it is arguably more difficult to make good silage than to produce hay of tolerable quality. Silage production involves the preservation of grass in a moist state by means of lactic acid, which is pro-

Fig. 7.6 A forage harvester cutting grass for direct transport to a silo

duced by a bacterium. If this process is to produce a nutritious food two conditions must be fulfilled: respiration of the grass must be halted as soon as possible after it is cut, and the bacterial fermentation must be carefully controlled.

After it is cut grass continues to respire and its temperature rises. If this is not controlled, the temperature rises rapidly to 40 °C, at which point its protein content is destroyed and its feeding value consequently much reduced. Control depends on the exclusion of oxygen from the cut grass, which must be quickly transported to a silo. This may simply be a rectangular concrete floor with high retaining walls on three sides (a *clamp* silo), or a specially constructed industrial tower. In both cases the grass is consolidated and covered to exclude oxygen, and if conditions are right fermentation will commence. If they are not right, lactic acid will not be produced fast enough, the temperature remains high, and other bacteria cause a butyric fermentation which causes the grass to rot. The failure to produce acid fast enough is due to a shortage of soluble carbohydrates, and to prevent this arising acid can be added or a source of carbohydrates, but both processes are difficult to carry out successfully. A surer alternative is to leave grass on the field to dry for several hours after cutting, and only then transport it to the silo. This *wilting* process reduces the moisture content, which increases the concentration of soluble carbohydrates. The grass may also be chopped into very small pieces of equal size, which allows the carbohydrate to leave the plant cells and thus speeds up fermentation. It also assists consolidation, which reduces respiration.

Fig. 7.7 Grass being compacted in an uncovered clamp silo. As it is filled the silage is covered with plastic sheeting to exclude air

Another major advantage of wilting is that in reducing the moisture content of the grass, it reduces the amount of liquid effluent which runs from the silo. Where a sealed tower is used, wilting is absolutely necessary since this effluent would not be able to drain away, but it is also advisable where a clamp silo is used. In the first place, nutrients are lost in the effluent, but more important, it constitutes a serious source of pollution, since a 400-tonne clamp of unwilted silage can produce as much pollution as the untreated sewage of a town of 150,000 inhabitants.

The farmer's basic choice in planning his winter feed is thus between hay and silage, but he may also purchase dried grass, or straw plus grain, and he may supplement the grass products with crops of swedes and turnips. The choice will depend on several factors, chief of which is the relative nutrient value of different foods related to their cost. It is possible to compare all the possible combinations in terms of the amount of usable energy and protein they provide an animal per hectare of land committed to their production. The comparison will also take account of the losses resulting from physical damage, weather conditions, fermentational spoilage, etc., and where purchased feed is also considered, its cost enters into the analysis (Table 7.1). This comparison is difficult but not impossible, but in many cases the farm's topography or factors like labour availability may override economic considerations. If the farm has an extensive grazing system with large fields (generally because of its topography), hay making is likely to be preferred. In an intensive situation with paddocks or strip grazing, where it is necessary for good management to cut some of the grazing area and high nitrogen dressings are applied, silage making is more likely. Where labour is short silage may also be preferred, because a well-managed silage system is

Table 7.1. Comparative costs of grass and non-grass feeds
(Davies, H., Welsh Agricultural College, Aberystwyth, 1981)

Feed source	Cost/tonne of dry matter (£)	Cost/mJ of energy (pence)	Cost/litre of milk (at 5 mJ a litre) (pence)
Grazing*	32	0.25	1.25
Home-produced silage*	56	0.56	2.80
Home-produced hay*	73	0.79	3.95
Purchased barley £105/tonne	121	1.00	5.00
Purchased hay £70/tonne	82	0.97	4.85
Purchased concentrates £150/tonne	172	1.32	6.60

* Inclusive of all costs such as fertilizer, contribution to reseeding cost, labour, machinery, etc.

marginally less labour-intensive at the cutting stage, and much less so at the feeding stage, since the entire system of feeding from tower silos can be mechanized (Fig. 8.4). Even where silage is kept in a clamp, cattle can also be allowed to feed themselves, which results in a substantial saving of labour, although it also results in some waste (Fig. 8.3).

Until comparatively recently silage was not commonly fed to cattle other than dairy cows, and rarely fed to sheep. One reason is that sheep and beef cattle are not generally housed in the winter, and it is easier to cart hay to them on the fields. Where they are housed on some lowland farms, silage is increasingly incorporated in their diet (Ch. 10). The other reason why sheep and beef cattle were rarely fed silage in the past is that their productivity is not as high as that of dairy cows, so it was not generally considered that their level of nutrition had to be as high. The cost of producing silage is not significantly higher than that of producing good hay, and it may even be slightly lower, but there is usually a higher investment in storage and feeding facilities. Cost is inevitably a major factor in determining the choice of hay rather than silage where the change to silage would involve capital expenditure. If a farm has old buildings, and capital is not available for convenient new buildings and silage storage, a hay-making system is almost certain to be retained. In other cases, however, convenience of operation may determine the choice. Where sheep and cattle are both kept on the same farm, for instance, hay-making is more convenient than making hay *and* silage, and will do for both.

The conservation system is thus related to the total farm system in which grass production is integrated, and there is almost as much variety in grass systems as there are individual farms. The following chapters necessarily reduce this variety to three basic farming systems which all rely to a greater or lesser extent on grass: lowland dairying, hill and mountain systems where grass is literally the only source of income, and mixed farms where grass is combined with arable production.

8

Dairy farming

In most developed countries milk production is the basis of the viability of most small farms, and an important factor in the utilization of grass on the larger units. For years it has also been one of the most profitable, widespread farming systems. In 1979/80 in the UK, for instance, if the net farm income per hectare for dairy farms is expressed as 100, the comparable figures for specialist cereals, arable and upland livestock farms were 40, 86 and 12. In the same year, some 63,100 farms in Britain had dairy herds, which represented 27 per cent of the total number of holdings. In 1955 the comparable figure was 141,010 dairy farms, constituting 38 per cent of all holdings. However, the average herd size in 1980 was 51 compared with 17.5 in 1955, while average production per head had increased from 3,000 litres in 1958 to 4,740 in 1980, and was still increasing at the rate of 2 per cent per cow per annum. Though the number of herds had therefore declined, production had continued to expand, a trend witnessed throughout the EEC.

Dairying is also one of the largest and most integrated farming/food production and distribution systems. Because milk is highly perishable, it needs fast and efficient collection, treatment and redistribution. In Britain this is coordinated by the Milk Marketing Boards (MMBs, see Ch. 12), and in other countries by producer cooperatives which range from very small local operations to large-scale national and even supra-national organizations. In this enterprise more than most others, the individual farm is thus dependent on and closely integrated in supra-farm operations and decisions which affect the long and short-term decisions farmers have to make. The dairy herd is also important as a major source of calves for the meat industry, accounting for over half the beef produced in the UK.

The major decisions to be made in determining a dairy system relate to calving policy, replacement policy, feeding practices, and milking and housing systems. The range of options for each has to be assessed on a given farm, in the light of the farmer's objectives. A farmer wishing to maximize profit may not choose the same system as one aiming for maximum yield. If a labour-saving system is the highest priority, this may entail reduced yields *and* reduced profits. In each case the farm environment will also shape the decisions. In a high rainfall, for instance, the feeding regime will generally rely heavily on grass, whereas in a drier region it may rely more on cereals.

CALVING POLICY

Cows can be made to calve when it suits the farmer best, but the three commonest patterns are spring, autumn and year-round calving. A number of
factors influences the choice, the easiest to assess being profitability. Liquid
milk consumption does not fluctuate much over the year, but supply does.
To ensure a stable supply to match demand, a premium is therefore paid for
winter milk, since production would normally fall due to higher production
costs (Figs. 8.1 and 8.2).

Farmers may therefore choose an autumn-calving herd which produces
most of its milk in the winter (since milk yield increases from calving until
the twelfth week, then gradually declines over the remaining thirty weeks of
the average lactation). Table 8.1 shows the variable costs of an autumn-
calving herd to be higher, reflecting the need to purchase feedingstuffs,
while fixed costs are also higher, perhaps reflecting the need to provide better and more expensive buildings. Even so, the gross output is still high
enough to offset these costs, yielding a higher return than spring-calving or
year-round herds.

Profitability may therefore appear to point to autumn calving, but farm-

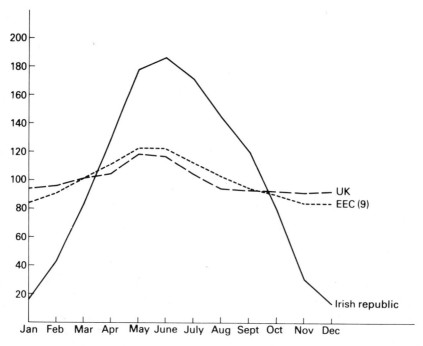

Fig. 8.1 Index of monthly milk deliveries in 1979 in the UK (Irish
Republic and rest of EEC included for comparison). Average daily supplies
in 1979 = 100
(Based on data in *EEC Dairy Facts and Figures*, Milk Marketing Board,
1980)

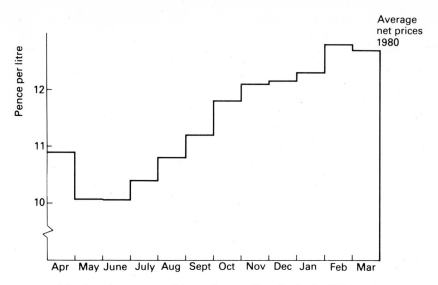

Fig. 8.2 Average monthly producer milk price in the UK in 1980 (Based on data in *UK Dairy Facts and Figures*, Milk Marketing Board, 1981)

Table 8.1. Cost and returns of different calving systems (per ha) 1976–77. (Jones, W. D. and Sherwood, A. M., *Dairy farming in Wales*, University College of Wales, Aberystwyth, 1980)

	Yield per cow (litres)	Gross output (£)	Variable costs (£)	Gross margin (£)	Fixed costs (£)	Management and investment income (£)
Spring calving	3,954	657	264	393	195	198
Autumn calving	4,941	903	355	548	235	313
Year-round calving	4,307	707	285	422	229	193

ers who want to make the fullest use of grassland may choose spring calving because it demands less purchased feed and lower capital investment in buildings. A family farm may, however, prefer year-round calving because it produces a regular income and has a fairly constant labour demand. From the point of view of capital investment it also has the advantage that equipment size (e.g. milk tanks) can be kept to a minimum and its utilization optimized.

HERD REPLACEMENT POLICY

The dairy farmer may breed his own replacement stock or purchase them from other farmers. In terms of simplicity the purchase of all replacements has much to commend it, and in strict economic terms it may be more profitable since none of the farm land is being used to rear unproductive stock. The validity of this argument nevertheless depends on the purchase price of replacements compared with the profitability of other uses which could be made of the land on which they would be reared. Where no replacements are reared, all cows are mated to a beef bull (e.g. Hereford, Charolais, Aberdeen Angus) to produce calves which command a higher price when sold for beef production (Ch. 10). The number of such 'flying herds' is probably a minority, however, and most farmers rear at least some replacements from their best cows (mated with a dairy bull) and cross the rest with a beef bull to produce beef stores.

The breeding and rearing of replacements to improve herd quality is a policy highly valued by farmers who practise it, and some who cannot rear replacements at home still arrange for home-bred calves to be reared on contract by another farmer. This allows for improvement within the herd, and simultaneously provides the rearer with a source of income which requires no capital of his own. Farmers breeding replacements select their best cows on the basis of conformation, milk yield or milk quality, and cross them with a bull with proven ability to correct defects and pass on improvements to his progeny. Natural mating with a bull on the farm is now uncommon, artificial insemination being the norm. (On only 9 per cent of dairy farms is natural mating used alone.) Artificial insemination services are offered by agencies which run cattle-breeding centres throughout the country where bulls of each breed are kept whose progeny have been shown to be superior. The records of these bulls are published, and a farmer can nominate one in particular or simply choose the 'bull of the day'.

Another advantage claimed for home rearing of replacements is disease control. Each farm has its own spectrum of diseases to which home-reared calves develop some immunity. Animals introduced from other farms may not have the same resistance, and may into the bargain introduce diseases to which the herd has no resistance. Another factor which may favour home rearing in some cases is farm layout. Many dairy farms have land which is too far from the buildings or is otherwise unsuitable for grazing by dairy cows, and it may be sensible to use it to rear heifers as replacements for the dairy herd, instead of grazing it for beef or lamb.

FEEDING REGIME

Like any other animal the cow obtains energy, protein, minerals and some vitamins from its diet. In nature the cow would obtain all its needs simply by grazing grass, but its milk yield would be low. In order to increase the yield farmers improve the supply of grass, and may supplement it with a concentrated feed compounded from cereals, proteins and minerals. The need for supplementary feed is obviously greater in the winter, but even in summer minerals may have to be fed, and concentrates provided for very high-yielding cows. A farmer wishing to rely on grass and conserved products to feed his herd has to have a spring-calving herd, with the result that yields will be lower, prices lower, and profits less. An autumn-calving herd relies more on purchased concentrates (or arable products on mixed farms) which produce higher yields and higher profits. The choice of feeding regime thus entails interrelated financial and nutritional decisions (which are further complicated by operational considerations like labour availability), and though there is clearly a minimum nutritional level, the optimum level for each cow will depend on its performance and the relative price of milk and feedingstuffs.

The nutritional aspects of the feeding regime are fairly well-established. Cows require energy and protein for four purposes: the maintenance of their essential life processes, their own growth, the growth of the foetus during pregnancy, and milk production. The energy and protein requirements for all these purposes have been carefully established, and are available to farmers in books, the farming press, and through their advisers. It is thus possible, for example, to calculate that a 500 kg Friesian cow, yielding 25 kg daily of 4.0 per cent butterfat and 8.6 per cent solids-not-fat milk at peak lactation, and losing 0.5 kg of weight per day, requires 179 mJ of energy per day and 1,730 g of protein. With knowledge like this the farmer (or his advisers) decide in what form to provide the necessary nutrition at minimum cost. This is an important consideration, since feed accounts for 60 per cent of the total cost of keeping a cow. Tables are available which provide in a ready-to-use form the energy and protein content of most available feedingstuffs, and the costs of these ingredients are also fairly easily obtained. Grass or conserved products are the cheapest source of nutrients, so most dairy farmers base their feeding on these (see Table 7.1).

At the start of the winter feeding period, a farmer who has silage available will assess its quantity and quality in terms of energy and protein, possibly having a dry-matter determination and a chemical analysis carried out on his behalf. He will know from experience how long this silage will have to last in an average winter, and after including a safety margin to allow for a longer winter than usual, he calculates the amount he can allow each cow. The energy and protein values of this can then be calculated and compared with the recommended feeding standards. For an autumn-calving herd this level of nutrition will be insufficient, so supplementary feeding will be needed. For a spring-calving herd, the silage plus some minerals may be adequate until a few weeks before calving, when supplementary concentrate rations may be provided. Proprietary feedingstuffs which contain all the necessary nutrients are readily available from manufacturers, and most farmers find this the simplest source of supplementary food. Others prefer to mix their own feed from cereals, protein and minerals, and they may thus succeed in making a nutritionally balanced ration at lower cost than manufactured feeds, especially if they can produce some of the inputs on the farm. In calculating the cost, however, they must not overlook the cost of the machinery and labour involved.

Deciding *what* to feed is only part of the art of feeding dairy cows. It has been established that the *level* of feeding affects future yield, and although many farmers determine the feeding level on the basis of current yield, others adopt a practice known as *lead-feeding*. The amount of feed is calculated in excess of current yield, in an effort to increase the milk yield at peak lactation, thus increasing the total lactation yield. In practice there are literally hundreds of different feeding systems adopted by farmers. Some are exceedingly simple, with all the cows calving at the same time and all being fed identically. At the other extreme farmers treat each cow as an individual. This method relies on very accurate recording of each cow's milk yield, on which the amount of food is directly based. Such cows are generally lead-fed according to a predicted yield, and groups of freshly calved cows are separated from the herd and fed separately. Almost certainly this method will produce very high individual yields on the basis of concentrate feeding, but a growing number of farmers argue that high yields are undesirable when there is already a milk surplus in the EEC, particularly when the milk is produced from high-priced concentrates. They argue that dairy farmers should rely more on grass, in which case yields would fall, but so would feed costs. However, as long as the milk price/feed cost ratio is high enough to make the production of milk from concentrates profitable, most farmers will not find this a persuasive argument, especially since many of them derive great personal satisfaction from breeding and feeding cows to achieve very high yields.

Even so, most farmers obviously try to reduce feeding costs by providing from their own resources as much forage and high-quality conserved material as possible, and by extending the grazing season as long as possible by growing special crops to supplement the grass. It was noted earlier that grasses require a minimum temperature before they start into growth, but that Italian ryegrass grows at lower temperatures than other grasses. The same is true of rye, so many farmers sow a mixture of rye and Italian ryegrass in August and September, which grows through the winter. In late January fertilizer is applied to provide early bite, and by the end of February the cows are allowed to strip-graze the crop behind an electric fence which is moved twice daily. When the grass begins to grow on other fields in late March, the cows are not simply left to graze at will, but are generally strip or paddock grazed throughout the summer (Ch. 7). By October, grazing will be insufficient and the cows are turned onto kale, a brassica which in some varieties produces thick stems high in starch, and in others large heads of leaves which are frost-hardy. The latter is particularly useful for post-Christmas feeding, or in areas with severe winters. This kale crop is sown in May after the rye has been ploughed in, and together with the rye it supplements the grass at either end of the grazing season.

Most of the cows' diet is composed of roughage – hay and silage, both of which are bulky and heavy in relation to their nutrient content. Feeding them has always been a highly labour-intensive operation, though as Chapter 7 noted, baled products like hay are easier to *feed* than silage, while hay

Fig. 8.3 Cows self-feeding at a silage clamp (Farmers Weekly)

making requires more labour and is more dependent on weather conditions. Consequently, as labour became more scarce and farmers recognized the advantages of silage as a conserved feed, new feeding techniques were developed (though some dairy farmers still make and feed hay). Silage feeding methods range from simple, low-cost self-feeding systems to sophisticated, high-cost fully automated ones. In the former the cows have 24-hour access to the silage clamp, the amount eaten being regulated by a barrier or electric fence which allows each cow on average to consume the desired amount (Fig. 8.3). Fully mechanized feeding systems have tower silos from which silage is periodically extracted by screw augers and delivered by conveyor belt direct to feeding troughs (Fig. 8.4). Between these two extremes there are methods which use tractors and trailers to collect silage from the tower or clamp and deliver it to the cows. These involve the use of labour, but are less prone to mechanical breakdown.

HOUSING AND MILKING ARRANGEMENTS

Farmers have several types of housing and milking systems from which to choose, and in many cases the choice has been shaped by the need to adopt new feeding methods, especially as silage has supplanted hay on many farms. In some climatic regions it would be possible to leave dairy cows outside throughout the year, but this is not generally possible in the UK. On some dry, sheltered farms spring-calving herds are sometimes out-wintered, in which case housing may be unnecessary, but milking arrangements are

Fig. 8.4 Mechanized silage feeding system (Farmers Weekly)

needed. In this situation the farmer may not invest in permanent milking premises, but use a *milking bale* – a low-cost mobile milking parlour. On the vast majority of farms it is nevertheless necessary to design a set of buildings which provide housing, feed storage and feeding accommodation in close proximity to the milking premises. This ensures an efficient, labour-saving unit, but the layout must allow the premises to be efficiently used in the summer when the cows are not housed. The complex therefore has to be sited close to the grazing areas, but also in such a position as to minimize the cost of providing all-weather access for heavy lorries delivering food-stuffs and collecting milk.

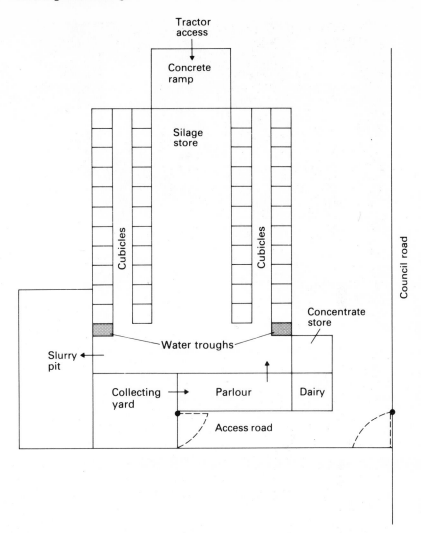

Fig. 8.5 Typical compact layout for dairy housing, milking and feed storage

When designing a new layout on a green-field site it is fairly easy to allow for these factors. In 1980, however, almost 50 per cent of British dairy farms (though only representing a small proportion of the national herd) were still using buildings erected many years ago which were designed to meet housing and milking requirements very different from those demanded by the most modern standards. The basis of the old system was a single cowshed in which the cows lived, were fed and milked (Fig. 8.6). Hay made on the farm was almost invariably used for winter feed and stored in close proximity in a dutch barn. This was an efficient system as long as herds were small, labour inexpensive and relatively abundant, and technology re-

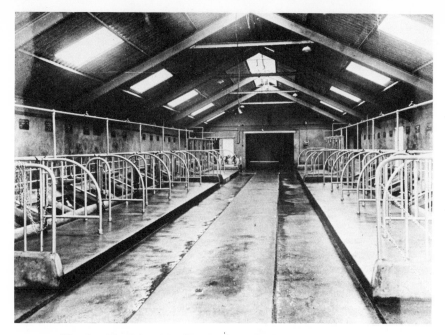

Fig. 8.6 Traditional double-range cowshed

latively unsophisticated. It was nonetheless a labour-intensive system. Hay was carried to the cowshed and fed in fixed mangers in front of the cows, which were tied and bedded on straw which had to be cleaned away before the twice-daily milking. Even using a modern milking machine (and it is worth noting that some 1,500 British dairy herds were still hand-milked in 1979) the milk was normally drawn into buckets which had to be manually carried to the dairy for filtering and cooling. The labour requirement fell with the advent of pipeline milking (which conveys the milk under vacuum to the dairy) and mechanical methods of dung removal, but the system was still labour-intensive and could not easily be adopted for the large herds established in the 1950s. Changes were also encouraged by the introduction of improved hygiene standards under which milk had to be produced (regulated by the Ministry of Agriculture).

The main effect of these changes was the separation into different buildings of the three functions of the traditional cowshed – feeding, housing and milking, as Fig. 8.5 demonstrates. This made the substitution of machinery for labour increasingly possible, which considerably lightened the tedium of a very repetitive enterprise. Even so the cowshed is still retained on many farms because capital investment in alternative systems is not judged worthwhile, or because farmers maintain that the frequent attention which the traditional cowshed entails results in better herd management. Where herds have increased substantially in size, however, the cowshed has been abandoned for parlour milking and separate housing.

Most cows are today milked in a single-purpose milking parlour. There are several basic types of parlour, each specially suited to a different herd size, but all allow for the same sequence of operations. The cows enter and are constrained in a *standing*; the udder is washed (a hygienic measure which also stimulates the let-down of milk); the fore-milk is scrutinized for signs of mastitis (a serious bacterial infection); the milking cluster is attached, and the milk drawn from the udder by means of a pulsating vacuum into measuring jars. The milk then flows from the jars into the refrigerated milk vat situated in the dairy. Immediately the milk flow ceases the teat cups must be removed to prevent damage to the udder tissue, and treatment to prevent infection may be given before the cows are released. When milking is over, all the equipment has to be sterilized and the parlour thoroughly washed down. This already time-consuming procedure is further complicated by the fact that while the cows are in the parlour they are also fed any necessary concentrates in their diet.

The advantage of parlour milking is that it increases the number of cows which can be milked per manhour, and this extra productivity is accompanied by a reduction in the hard work involved. Increased labour productivity during the milking operation is limited, however, by the amount of time necessary for the milk to be extracted from the udder, which is approximately 5–7 minutes. This cannot be speeded up, so any savings must be made on the associated activities. Another limiting factor is the time it takes a cow to eat any food which is provided. A cow can eat about 2 kg of cubed concentrates in the time it takes to be milked. The feeding of more than this amount would therefore reduce the number of animals which can be milked through a parlour of a given size in a given time. This problem does not

Fig. 8.7 Herringbone milking parlour in use (cf. plan Fig. 8.8a)

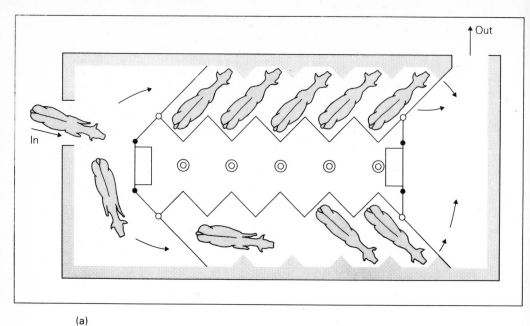

(a)

(b)

Fig. 8.8 Two commonly-use designs of milking parlour
(a) herringbone
(b) rotary herringbone (Longman Group Ltd. from *Lactation of the Dairy Cow*. C. Whittemore, 1980)

arise for spring-calving herds during the summer, since there is no need to feed large quantities of concentrates, but for autumn-calving herds the problem can be serious. One solution is to feed a basic production ration to all the cows by putting troughs in the yard, and supplement this at milking time on an individual basis. This system is satisfactory when all the cows calve within a fairly short period. Other systems involve the division of the herd into various yield groups which are fed separately, but this means that time is saved in the parlour and feeding efficiency increased *only* at the expense of increasingly complex management. A farmer in this situation therefore has to decide whether or not the benefits involved are greater than the costs.

Different parlour designs allow different numbers of cows to be milked per hour. One of the commonest designs is the herringbone parlour, which has a throughput of 80 cows per manhour. Rotary parlours in which the cows stand on a platform which rotates around the cowman have a throughput of 90 per manhour, and recently introduced polygonal parlours may further increase the average time available to milk a cow without reducing the throughput. Electronic devices are also used now to speed up the work routine. Cows may, for example, be fitted with a radio transmitter collar which identifies the cow and switches on an automatic feeding system which delivers her correct ration. Automatic removal of the milking cluster is also increasingly used, and prevents over-milking.

Once the milking accommodation was separated from housing and feeding accommodation, it became possible to re-design the last two with a view to reducing the cost per cow housed and the labour requirement for feeding and cleaning. The basic need is for comfortable, well-ventilated accommodation which also keeps the cows as clean as possible (since this reduces the time spent on udder washing in preparation for milking). One solution is *loose housing*, in which the cows are kept in covered yards on a layer of straw. This may suit arable farms where plenty of straw is available and farmyard manure is needed for application to crops. These yards are cleaned out once or twice a year, so they reduce very substantially the labour required for the traditional cowshed. If the cows are to be kept clean, labour is nevertheless required to spread fresh straw daily. This increases the labour requirement at this stage, but may be worth it to save time during milking.

On dairy farms without arable enterprises the purchase of straw for bedding is an additional cost, whereas its use on an arable farm is a convenient way of converting a useless by-product into a valuable soil conditioner and fertilizer. An alternative to loose housing is therefore needed which calls for relatively little bedding material, which leaves the cows untied so that they may self-feed, and is relatively cheap and easy to install in existing or new buildings. The commonest solution is *cubicle* housing (Fig. 8.9) which provides stalls in which the cows can lie down but are not tied, so that they

Fig. 8.9 Cubicle housing (Farmers Weekly)

may come and go to feed at will. Only the floor area immediately under the cow is covered with sawdust, chopped straw or a rubber mat, to provide comfort without incurring high bedding costs. The cubicle length is carefully regulated to ensure that dung and urine cannot fall in the cubicle, but into the wide passage separating the rows of cubicles. This arrangement facilitates cleaning and is also low in capital cost, but it has one major drawback – the disposal of slurry.

Straw-bedded loose housing has the advantage that dung and urine are absorbed by the bedding material, so the resulting manure can be handled as a solid. Cubicles inevitably produce a fluid manure which is difficult to manage, though it is possible to handle it as a semi-solid, using a tractor and low-cost muck-spreader. Alternatively, when the manure has been mechanically scraped from the cubicle house, water is added (or rainwater is allowed to mix with it) to produce a liquid which can be stored in a tank or pit and applied to the land when weather conditions permit. Attempts have been made to spray this slurry, but machinery blockage and problems of drift caused this technique to be very largely abandoned, certainly where proximity to residential areas brought complaints from the public. Even though the technical problems have been effectively overcome, slurry is therefore generally applied as a liquid from a tanker, thus reducing the problems of drift. On some farms a pit is excavated under the passage between the cubicles and covered with a slatted floor which eliminates the need to scrape manure mechanically. The manure simply falls into the pit and flows under gravity as required into a mechanical spreader. The cost of such a system is initially high (except perhaps on a sloping site), but it saves labour and machinery costs, and significantly eases a laborious routine.

Fig. 8.10 Slurry handled as a semi-solid, stored in an open pit

Fig. 8.11 Liquid slurry pumped into an above-ground storage tank

SUMMARY

The commonest dairy system was long based on a self-contained herd of approximately 25 cows, housed and milked in a cowshed, fed on hay and some concentrates in winter and set-stocked in summer. As herd sizes grew, cubicle housing and parlour milking progressively replaced the cowshed, and cows were self-fed on silage and more concentrates in winter, and paddock or strip grazed in summer. Either system is viable for spring- or autumn-calving herds, though the latter has consistently proved more profitable because of the high price traditionally paid for winter milk. As supply has increased, this price differential has been reduced, and the oversupply of milk has led some farmers to argue that the emphasis on winter production achieved by high concentrate feeding is wrong. They maintain that the efficient dairy farm should be an efficient machine for the conversion of grass into milk, so the emphasis should be transferred to spring-calving herds relying mainly on grass and better grassland management. However, such a move towards a low-output/low-cost system is unlikely as long as the ratio of milk price to concentrate price maintains the profitability of high-output/high-cost systems.

SPECIAL TOPIC: THE OVERSUPPLY OF DAIRY PRODUCTS IN THE EEC

The dependence of the individual farm system on factors beyond the farm-gate is well illustrated by the problem caused by the oversupply of dairy products in the EEC. This surplus is the creation of the Common Agricultural Policy (Ch. 3) whose object was deliberately to support farm incomes as a social measure in the original member countries. The foundation of many small farms' incomes in these countries, as in the UK, is milk production, and the entire preoccupation of the efficient dairyman is to increase milk yield or reduce costs in order to earn the highest possible return for the considerable effort and capital committed to the enterprise. The problem is that it is easier to increase milk yield than it is to reduce costs, so the irony is that the efficient dairy farmer is contributing to the enormous surplus of milk produce which the tax system of the EEC has to subsidize. It would be difficult to find a more obvious example of the conflict between the farm system, well-integrated and perfectly rational on its own level, and wider public interests at another level of integration.

Research suggests that unit costs are lowest at a herd size in excess of 100 cows, and figures published by the MMB of England and Wales showed that in 1978/79 dairy farms in this category had the highest gross margin per cow and per hectare. Since the present average herd size in Britain is only 51 (and is considerably lower in other EEC countries), cost-conscious farmers are tempted to become more efficient by increasing their herds accordingly. They have been encouraged in this policy by successive governments because of the contribution which domestic milk production can make to import saving or to export earnings, and thus to the balance of payments. In France, the dairy industry is one of the country's biggest export earners. In Britain the industry is less an exporter than a saver of imports. In 1979 the UK *imported* £300 million of butter, £190 million of cheese and £38 million of other milk products, while it *exported* only £226 million of milk products, leaving a dairy sector adverse trade balance of £302 million. It has been argued that expansion of the national herd would reduce this figure substantially, and when the costs are neglected this is a plausible argument. However, extra production would have to come from purchased feed – much of which, especially the protein content, has to be imported. This import cost therefore has to be offset against the import saving from increased domestic production. The saving has nevertheless been substantial enough to persuade governments to encourage the expansion of the dairy herd even though they have simultaneously criticized the dairy surplus at the European level.

The reasoning behind this apparent inconsistency is the argument that the production of specific commodities should be encouraged where the optimal natural conditions and farm structure for that production exist.

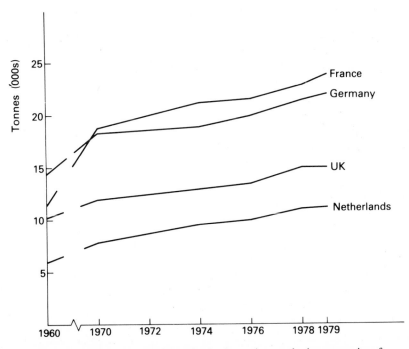

Fig. 8.12 Expansion of milk deliveries in the major producing countries of the EEC 1960–79. (Total EEC production in the same period increased from 57,315,000 to 93,315,000 tonnes)
(Based on data in *EEC Dairy Facts and Figures*)

Such is the case, it is argued, for milk production in the UK. In strictly agricultural terms the argument is difficult to fault, for British dairy farmers compare well in terms of productivity and efficiency with the best in Europe; the climate is better suited to grassland production than it is in many other European countries, and the farm structure is suited to the achievement of the economies of scale of large herds. The difficulty is that if dairy farming is the mainstay of small farms in this country, it is even more so of the very small farms which exist in other member countries, and the argument for agricultural efficiency would eliminate most of these uneconomic dairy units. Even though, as Chapter 3 noted, many of these small farms are not full-time holdings, such a move would create serious social and political problems in countries like France and West Germany. Politicians in these countries cannot afford to disregard these problems because the agricultural vote is still important enough to determine the result of local and national elections. By the same standard of agricultural efficiency, moreover, British farmers in some enterprises which are uneconomic compared with the same enterprises in Europe (e.g. sugar beet and glasshouse production) could also be argued out of their livelihood. Any British government would find this difficult to accept on grounds of social justice, if not out of fear of the electoral power of the agricultural population.

The problem is further complicated by the critical fact that the consumption of dairy products is at best constant, and there is some evidence that it is declining over the long term. There seems little real possibility of reversing this trend unless consumption patterns change radically, which would mean changing eating habits that are very well established. This is therefore no easy solution to the growing dairy surplus. On the contrary, resistance to price rises and fears about possible health risks associated with high consumption of animal fats have apparently consolidated the position in which stagnant demand is combined with ever-rising production levels. Britain's co-members of the EEC have always believed that the solution is to exclude third-country imports into the UK, primarily from New Zealand, filling the gap with EEC produce. Special arrangements were negotiated by Britain before accession to the Community which allowed New Zealand imports until the end of 1977, and this dispensation was subsequently extended to allow fixed amounts of butter imports until the end of 1982. Given the growing butter mountain, however, the government has come under increasing pressure from British farmers as well as other members of the EEC (especially Ireland, which sees Britain as its natural market) to negotiate no further extension of the New Zealand arrangement. This is a sensitive political issue, since there are longstanding ties with this Commonwealth country whose economy could still be harmed by a ban on exports to the UK, even though it has made great efforts in the last decade to diversify its markets to reduce its former dependence on the British market. The British government's resistance to pressure to exclude New Zealand produce has led to calls from the EEC to curtail UK production, which the government finds equally unpalatable because of its political implications.

Even this very limited outline of the problem underlines the difficulty of finding a solution to the oversupply of dairy products which is capable of satisfying all the competing interests. British and European farmers wish to expand production to increase or maintain their incomes. The British government would like to expand domestic production in order to save imports and improve the balance of trade. New Zealand farmers and their government wish to avoid a restriction of their market and consequent loss of revenue. Irish farmers and their government want to increase their share of the British market in order to raise farm incomes, and the EEC Commission has always seen the exclusion of New Zealand imports as the easy solution to an intractable problem which member governments have proved consistently unwilling to confront directly by cutting the milk price dramatically or introducing production quotas.

The result has been a history of compromises, involving at the EEC level taxes on over-production levied on farmers, attempts at a price freeze, payments to producers to transfer from milk to beef production (even though there is also periodically a beef surplus in the Community), and active consideration of a quota system. At the level of individual member countries, including the UK, governments have urged farmers to increase efficiency and improve their marketing tactics to secure a larger share of the market. The British government has simultaneously deplored the milk surplus and continued to pay grants to improve or expand dairy units, and European governments have used similar national policies which counteract every effort to tackle the problem of oversupply at the supra-national level. Through it all, individual farmers have made production decisions as rationally as circumstances allow, finding themselves criticized if they do *not* increase their efficiency *and* if they increase it such that the surplus is increased yet further. The example demonstrates only too plainly how the external environment conditions the individual farm system, and how intractable the situation can become when the issues involved are less agricultural than social, political and economic.

9

Grassland systems in the uplands

A third of Britain's agricultural area is classed as *upland*, of which a high proportion is classed as rough grazings. The farming systems in these areas fall into two main categories: true mountain farms with no low land near the farmstead, and upland (*marginal*) farms where the farm buildings are often located in a valley, surrounded by some relatively level fields (the *in-bye*), while the rest of the land is held as hill and mountain grazings. These mountain grazings may be several miles from the farmstead, which considerably complicates management and may prevent altogether some desirable practices like rotational grazing. True mountain farms are confined to northern Scotland, the northern Pennines and North Wales. The land tenure pattern of these mountains is complicated. Sometimes they are held in common, especially in northern England, so it is impossible to fence them. They are sometimes owned by large landowners whose principal interest in them is to raise game for shooting, so there is again no possibility of enclosing the land, and the grazing system is necessarily extensive. Where farmers own their mountain grazings it may be possible to fence all or part of the holding, but the system is still extensive, and almost exclusively restricted to sheep production.

The ownership pattern of marginal farms can be equally complicated in some regions of the country, but what mountain and marginal farms certainly have in common is the harshness of the natural environment, which severely limits the range of possible farm enterprises. The short growing season and high rainfall make arable crop production impossible, so the entire system is based on grass, and even this may be poorly utilized. The stocking level generally has to be restricted to the number of animals for which winter feed can be provided, which in most cases is low, and sometimes very low. The type of stock kept must also be chosen for their hardiness, not their productiveness, and these conditions together mean that upland farm incomes tend to be low.

The British government traditionally tackled this problem through various schemes which grant-aided hill improvement, with the intention of increasing the productive potential of the hills and therefore farmers' incomes. The EEC has a similar policy, since the problem of marginal farms is even greater on the Continent, but the policy is increasingly questioned in

Britain by the conservation lobby. Instead of grants for agricultural improvement, conservation interests argue that income supplements should be paid to farmers in compensation for the loss of income from improved pastures. What is not generally appreciated is that such a policy would have repercussions for the entire meat industry, since upland farms produce the store stock which is fattened, or used for breeding other stock for fattening on lowland farms. Some 70 per cent of all ewes and 55 per cent of beef cows in Britain are produced in the uplands. The number of store stock which upland farms can supply for the meat trade is limited by the amount of winter feed they can provide, which we suggested is often very little. It is possible to purchase feed from lowland farms (Ch. 2), but this increases the production cost and squeezes hill farmers' margins. Moreover, less feed is now available from lowland farms, so unless hill farmers continue to improve upland pastures to increase their winter keep, store stock numbers could decline. This would be reflected in higher livestock prices, leading almost certainly to higher meat prices. So the problem is not just one of supporting farm incomes in the marginal areas; it is also a problem of increasing food prices to consumers. The conservation lobby may well be prepared for this contingency, but the evidence is that it does not represent the views of the majority of the population, for whom meat prices loom larger than concern for the natural environment, or indeed the welfare of farm livestock (Ch. 11).

Unimproved upland grazings are composed of heathers, bilberry, low-yielding hill grasses which tolerate harsh conditions, and, on drier slopes, bracken. All of these make poor pasture, and since they are usually unfenced and badly drained, improvement in most cases is limited to periodic burning, which ensures a flush of new, more nutritious growth. A grass-dominant hill can sometimes be improved also by the application of lime and fertilizer, but even then it will not provide year-round grazing. Needless to say, none of these will provide any surplus for conservation. On marginal farms with some in-bye, some conserved forage and root crops for winter feed can generally be produced, and the in-bye can also be improved more easily since it is invariably fenced and can be drained. The size of the in-bye is thus the main factor determining stock numbers on marginal farms, and the lack of it is the severest constraint on real mountain systems. Where provision cannot be made on the farm for winter feed, it has to be bought in, or stock must be sold off in the autumn or wintered away on other farms, thus allowing the maximum use of summer grazings while reducing the number of mouths to be fed through the winter.

On true mountain farms (which have as little as 15 per cent of their land fenced) the vast majority of the stock are sheep, though some cattle may be kept on lower slopes in more favoured situations. The proportion of sheep to cattle on such farms is in the order of twenty to one. On marginal farms more cattle will be kept, the numbers of sheep and cattle being determined by the ratio of hill to in-bye, and the existence on the farm of fenced im-

Fig. 9.1 A typical section of an upland farm, showing sheep on the in-bye, and an unimproved bracken slope behind

proved mountain. If 70 per cent of the farm is fenced, there may be one cow for every five ewes. Each of the hill areas has its own breed of sheep and cattle, and there has been remarkably little inter-breeding. The principal cattle breeds are the Highland, Welsh Black and Galloway, but there are many more breeds of sheep, whose names indicate their origin – Scottish Black-

Fig. 9.2 A pure-bred Scottish Blackface flock brought down before drafting (Farmers Weekly)

face, Welsh Mountain, Swaledale, Derbyshire Gritstone, etc. Each is stoutly defended as the only breed which could possibly survive in local conditions, but this is not strictly true. All the hill breeds produce hardy ewes which make good mothers and are well-adapted to their environment, but none of which are very productive. Farmers over the generations have tried to improve productivity by cross-breeding, not just with other hill breeds, but with lowland breeds which produce a bigger meat carcase and have a higher lambing percentage (produce more lambs per ewe). Since the cross-breeds are generally larger, however, and their food requirements higher, any improvements resulting from cross-breeding will be lost through poor nutrition if food supplies cannot be increased, so fewer ewes will have to be kept. Improvements in breeding and feeding therefore go hand in hand, and the argument is unending as to whether it is better to keep a smaller number of bigger ewes, or a larger number of small ewes. The argument has been influenced in Britain by the subsidy traditionally paid to hill sheep producers on a headage basis. This has unquestionably been a major disincentive to improvement of the quality of upland sheep, though the difficulties of achieving improvements in the harsh conditions of many farms should not be underestimated.

MOUNTAIN SYSTEMS

The simplest hill farming system, practised where the hardest conditions prevail, is the rearing of store sheep and cattle. The breeding stock produce their progeny in spring; the young remain with their mothers until weaning in late summer, and are sold in the autumn. At this stage they are not fit for slaughter (not *finished*). The cattle are too small and have to be fed for at least an extra year before being slaughtered, which is impossible on the farm of origin. The young cattle (*suckled calves*) are therefore sold to lowland farms for finishing. The ram lambs will also not have reached slaughter weight or condition, but in their case finishing may take only 4–12 weeks under the improved conditions of the lowlands. The ewe lambs are kept for breeding in the home flock. (For the links with lowland finishing systems, see Chapter 10.)

The rationale behind this system is that there is insufficient grazing available by late summer to bring lambs into finished condition, and insufficient winter feed to carry calves over the winter in addition to the breeding stock. Hill farmers are also traditionally reluctant to provide supplementary winter feed for *any* of their stock. This is due less to the high cost of purchased feed than to their belief that winter feeding undermines the hardiness of hill stock. However, this means that summer grazings are under-utilized because stock numbers are limited to those which will survive unaided on winter grazings. If farmers would feed more in winter, the overall utilization and profitability of mountain farms could therefore be improved, but the cost of feed reinforces their faith in the natural capacity of mountain stock to survive from their own resources. This pattern could be altered by the wider use of feedblocks, which provide protein, minerals and energy in a convenient form which lasts much longer than alternative feed. A 25 kg feedblock will last 30–40 sheep for a week, supplying up to 115 g/head/day of highly nutritious food, whereas any hay fed to sheep has usually been of poor quality. By providing sufficient blocks for the sheep numbers on the mountain a food supply is thus readily available, and this overcomes another unspoken objection to winter feeding, which is that transporting hay over mountain terrain at regular intervals in the depths of winter is not a pleasant job. The cost will nevertheless remain a serious disincentive, for it is twice that of other purchased compound feedingstuffs (per unit of energy), and since many farmers never feed anything more than hay, this cost seems astronomic.

One way in which stock numbers can be reduced in the winter is by *tack* or *agistment*. Ewes are normally put to the ram in October/November, when that year's crop of lambs is only six months old, and too small for successful breeding. If these unproductive ewe lambs are overwintered on the home farm they consume food which would be better used to feed in-lamb ewes. Traditionally they have therefore been sent to lower farms where they graze grass not required for other stock. Dairy farms were the main providers of agistment or tack accommodation, as it is called, for which the upland farmer paid a fee which was a useful addition to the dairyman's income. At the same time, his pastures were kept in good grazing condition over the winter, when it was impossible to turn cows out onto the wet land. However, dairy farmers wishing to increase the productivity of their land are increasingly reluctant to provide tack for upland sheep, because their grazing prevents the production of early bite for the cows. The reduction in the availability of tack accommodation, and its high cost when it is available, have therefore compelled many hill farmers to reclaim land on the home farm to provide better winter grazing or more conserved forage, or to house ewe lambs and ewes in buildings where they can be fed over the winter, with inevitable effects on the upland landscape.

The harsh conditions of mountain farms have prevented any opportunity for livestock improvement, and many observers argue that the quality of some hill sheep flocks is actually declining. The sheep enterprise on these farms is based on the self-contained flock, and only rams are bought in. One reason is that although the sheep graze open hills, the flock from one farm invariably grazes the same piece of land – called the *heft*, and lambs learn from their mothers to keep to this area. The hefted flock thus has a regular territory and is generally sold along with the farm, since a new flock would wander and not know where to find shelter in severe weather. The same is also true of bought-in replacements which might be used to improve the flock, however, so what is in one sense an advantage is also a constraint on hill flock improvement. The possibility of improved breeding is also reduced by the practice of selling ewes from hill flocks at only four years old, when they are considered too old to continue to thrive in upland conditions. These ewes are sold not for slaughter, but as *draft ewes* which go for further breeding on lowland farms where they respond well to the improved conditions, often improving their lambing percentage considerably.

The problem may be illustrated by a flock of 100 breeding ewes, of which a third will need replacing each year as 4-year olds are drafted out (Fig. 9.3). In addition there may be losses of 10 per cent resulting from severe weather, and for the same reason only 80 lambs may survive. Roughly half of these will be ram lambs, which are sold as stores to lowland farms. The 40 ewe lambs will be needed to sustain the breeding flock, and there will be no surplus to cover disasters, and no possibility of culling poor anim-

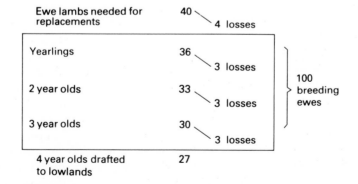

Fig. 9.3

als or of selecting stock systematically to improve the flock, whose performance inevitably declines.

The only sales from such a flock are the 27 draft ewes, the 40 ram lambs and the wool, which constitutes about 15 per cent of the total revenue. One of the only ways of increasing revenue might be to produce as much wool as possible of as high a quality as possible. However, the first job of the fleece is to protect the sheep against rain and cold, and for this purpose a lightweight fleece composed of two layers is necessary, the inner layer of soft, close fine wool being protected by an outer layer of long, coarser fibres which shed water. Such a fleece is not of the highest quality as measured by fineness and length of staple. Nor will it be of great weight, since protein is necessary to produce a heavy fleece, and in hill areas all the protein available to a ewe is needed to sustain her own growth, the growth of the foetus during pregnancy, and the production of milk to feed the lamb after birth. The farmer thus has to compromise between the desire for a higher income from wool and higher costs in sheep losses or increased feed to sustain the ewes, and in the end it is better that 100 ewes survive, even though they produce only 1½ kg of wool each, than that 80 ewes survive which produce fleeces of 2 kg weight.

On the true mountain farm, little can be done to improve the farm system. Productivity is therefore very low per hectare, and to make a reasonable living large areas of land are required, stocked at a rate of approximately one ewe to two or three hectares. The in-bye is often so limited that ewes are left on the mountain for most of the winter, grazing what little grass is available, and when this is snow-covered they survive by eating the green shoots of heather. To keep heather fit for grazing, it must be kept reasonably young and short, which is achieved by burning roughly every ten years. The timing of burning is regulated by law in such a way as to reduce damage to wildlife, and it is also generally controlled by the landlords who own a high proportion of mountain grazings, which are also let for shooting and thus shelter a large population of game birds which must not be harmed. The burning of the grazings is equally beneficial to the game (particularly grouse) and other wildlife which feeds on heather, since they too need young vegetation. There is therefore a common interest between farmer and sporting landlord which ensures that the operation is carried out under very closely controlled conditions. Where it is not, active education rather than statutory regulation is likely to achieve more in remote areas difficult to keep under surveillance.

MARGINAL FARM SYSTEMS

The structural and natural difficulties of marginal farms are less severe than those of mountain farms, and the farm system is more flexible. The in-bye generally represents a much higher proportion of the farm hectarage. There may also be an area of fenced, improved or unimproved hill, and an area of rough grazing which is often part of the farm, and can therefore be improved. The range of enterprises possible on such farms is consequently wider. Sheep and cattle production are still the mainstay of the farm business, but the cattle enterprise is likely to be more substantial. In favoured areas many farms have a dairy enterprise, and some may be in seed-potato production, but since these were discussed in earlier chapters, the account here is confined to the sheep and cattle systems.

The obvious way to increase the productivity of the sheep system is to finish lambs on the farm. Another way to increase farm revenue is to keep ewes longer before drafting, since it is well-established that their fertility increases with age. The 4-year olds drafted from mountain farms are only just reaching their most productive stage, and though mountain farmers believe they will not survive longer under their conditions, there is no reason why they should not be kept longer on lower farms. In both cases, however, the first priority must be to improve the pastures and the general level of nutrition of both ewes and lambs. This should also allow larger, more prolific sheep to be kept, and the higher level of nutrition should increase the lambing percentage. The critical period of the year is the two months immediately before lambing, when extra feed is needed to sustain foetal growth, and to reduce the cost of purchased feed most marginal farms grow crops of swedes on the in-bye for feeding after Christmas. The flock is still generally a pure-bred mountain breed, but ewe lambs are usually in sufficient surplus to allow selective improvement on the grounds of size and prolificacy, as well as adding to the revenue from ram-lamb sales.

A more complicated system which is commonly practised is to run two flocks on the same farm. One is treated as a pure-bred flock to produce replacement ewes hardy enough to graze the hill slopes. The second flock is composed of the older ewes which, instead of being drafted, are crossed either with a fat-lamb breed like the Suffolk, which produces a fine meat carcase, or with a breed which produces highly prolific ewes which can be sold for further cross-breeding on lowland farms. The breed most commonly used for this purpose is the Border Leicester. The crosses which it produces are called *half-breds*, which constitute the best fat-lamb flocks in Britain, and their local names reflect the particular cross involved (e.g. Welsh Half-bred = Border Leicester × Welsh Mountain). In this system no store stock is produced on the farm, and only the best ewe lambs are retained to maintain the breeding flock. The system also increases substantially the revenue from upland sheep production, and by providing highly prolific cross-bred ewes for crossing with fat-lamb sires on the lowlands, it is the basis of the entire sheepmeat industry (Fig. 9.4).

The cattle system on marginal farms is generally geared to the production of suckled calves weighing 230–280 kg when they are sold in October for finishing on lowland farms. Sometimes a proportion of the calves are kept for finishing on the home farm during the winter, or on grass the following summer. On most farms all the cattle are housed in the winter. The cows

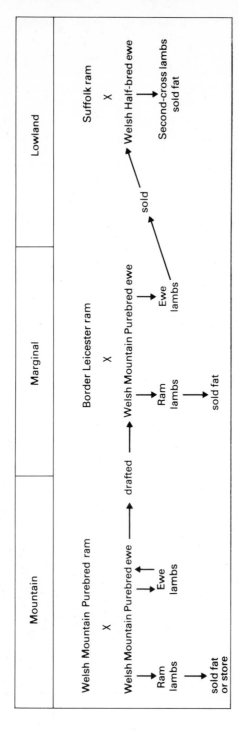

Fig. 9.4 Stratification of the British sheep industry, showing the interrelated nature of mountain, marginal and lowland sheep systems

Fig. 9.5 A pure-bred Border Leicester ram, used on mountain ewes to produce half-breds(Photograph: Douglas Low)

Fig. 9.6 Welsh Halfbreds produced by crossing pure-bred Welsh Mountain ewes with a Border Leicester ram

are fed mainly on conserved grass (generally hay), supplemented by a small amount of purchased concentrates. The calves being finished for the spring market are also fed sufficient hay and concentrates to achieve a daily liveweight gain of 1 kg. The calves kept in store condition for fattening the following summer are fed hay and very little else, so that they gain about 0.5 kg/day.

The most important decision relating to the management of the suckler herd is the calving season. Autumn calving produces a maturer and heavier calf which is more saleable the following October, but the cows have to be fed better through the winter to produce sufficient milk for the calves and to maintain their own body condition sufficiently well to ensure a good conception rate. Autumn calving is therefore a fairly high-cost system. Spring calving, on the other hand, tends to have lower costs since the peak feed demand occurs at the same time as the peak of grass growth. It is generally believed that spring-born calves fetch a lower price, however, though there is in fact little difference between the gross margins for the two systems.

One important factor affecting the gross margin is the regularity with which cows calve. Ideally each cow should produce one calf per year, which means that it must come into oestrus and conceive again within three months of calving. Often this does not happen and *slippage* occurs, the period between each calving becoming increasingly long, so that autumn-calving herds may eventually become spring-calving, the calvings being spread over a six-month period. One reason for this is poor nutrition, resulting in inferior condition and failure to conceive, so a spring-calving herd is in this respect easier to manage, since the grass growth should provide enough food to improve the condition of all but the worst-fed cattle. In the autumn-calving herd, where mating takes place in mid-winter, the farmer has to provide sufficient feed to maintain this condition, and on a hill farm this is not easy since winter feed is invariably in short supply.

The type of cow used for suckled-calf production varies with the severity of the environment and the region of the country. On the highest hills pure-bred Highland, Galloway or Welsh Black cattle may be used, but it is commoner to use cross-bred cows. In the harsher situations the Blue-Grey is often used (White Shorthorn sire × Galloway dam), and on better farms the Hereford × Friesian (which has also moved into increasingly harsher situations in recent years). These cross-bred cows are mated with a third breed to exploit the advantages of hybrid vigour. Calves of such a crossing programme survive and grow better, and may have an advantage of up to 25 per cent in weaning weight. The breeds of bull used are those which produce high-quality lean beef at a fast rate of gain – South Devon, Hereford or Aberdeen Angus, but recently larger foreign breeds like Limousin and Charolais have been used. The problem with these bigger bulls is that there is an increased risk of difficult calvings, and the interval between calvings tends to be slightly longer than when a Hereford or other smaller home breed is used.

Cattle on hill farms are generally rotationally grazed at fairly low intensity. They are not commonly out on the hill for long, even though this can benefit the grazings since they eat the longer grasses which sheep cannot use. Poaching of the land may nevertheless offset any improvement to the grass, and the cattle do not benefit from the experience. On most farms each cow produces only one calf, which is itself a low-intensity system, and since she has the capacity to raise two, some farmers purchase extra calves to suckle with home-born calves. *Multiple suckling* is an obvious way of increasing revenue, but it is not widely practised because it is often difficult to persuade the cow to act as foster mother to a second calf. The financial viability of the system also depends on the price of purchased calves, which in some years can be as high as the expected return from their sale six months later. Farmers therefore tend to act as opportunists, buying calves if they are cheap and adopting multiple suckling with some of the herd. If the calf price is too high, a single-suckling system is retained.

The combination of sheep and cattle production on the marginal farm is fairly easily integrated, with very few internal conflicts. While the pure-bred sheep graze the hills in winter the older ewes used for fat-lamb production graze the in-bye, while the cattle are housed. The sheep on the in-bye can be fed supplementary rations or graze specially grown crops of roots or kale, which improves their condition for lambing and produces more and stronger lambs. The higher level of nutrition also produces more milk to feed the lambs, so their growth rate is faster, and some may be sold before lamb prices fall sharply as the season progresses. After lambing these sheep can be moved to part of the in-bye which has not been grazed, or onto a sheltered piece of hill along with the cattle, while the pure-bred flock remains on the hills. The in-bye can then be fertilized and allowed to produce a hay crop or roots for winter feeding. An outline of the operations involved is given in Table 9.1.

HILL IMPROVEMENT

The harsh natural conditions of the uplands has always meant low stocking intensity, and since farms are generally small, incomes have also been low. In all countries where such conditions prevail governments have taken measures to support farm incomes, occasionally through income supplements, but in Britain the incentives have been towards the improvement of hill grazings, on which the increased productivity of the hills and the livelihood of hill farmers depend.

Hill improvement involves four interrelated processes: drainage, fencing, renovation or reclamation of grazings, and utilization. With the exception of steep hill slopes, a high proportion of upland grazings is very wet. It may consist of quaking peat bog or rushy grassland, and some drainage is an essential pre-requisite for improvement of the grazings. Since the return on the investment is low, the drainage must be inexpensive. On lower land

Table 9.1. A hill farming calendar

	Sheep flock	Suckler herd
November–Xmas	On improved hill.	On in-bye, fed outside on hay.
Xmas–March	Fat-lamb flock on in-bye, fed hay and concentrates, lambs March–April. Purebred flock on hill, lambs on in-bye late March–April.	Inside, fed on hay/silage and concentrates. Calves January–March.
April–May	Ewes and lambs on improved hill.	Cows and calves on in-bye.
May–June	Purebred ewes with single lambs to mountain. All others on improved hill.	Cattle to improved hill. In-bye closed up for hay/silage.
July	Ewes down for shearing. Lambs weaned and put onto hay and silage aftermaths. All ewes to mountain.	Bull run with cows on in-bye.
August–September	Ewes for drafting as 3–4 year olds selected and sold. Sales of finished lambs begin.	Cows and calves on in-bye.
October	All ewes on improved hill or in-bye. Ram put in after two weeks. Ewe lambs sent away on tack.	Calves fed some concentrates, then sold.
November	Ram removed. Purebred ewes to improved hill, fat-lamb flock on part of in-bye.	Cows on in-bye.

drainage generally involves the insertion of underground pipes covered with gravel and 30 cm topsoil, but this costs about £600 per hectare, so in the uplands it is out of the question. The solution here is open drains, ploughed with a forestry plough or excavated by mechanical digger, the object being to intercept springs and carry their water to rain channels and existing natural streams. There will nonetheless be areas which it is impossible to drain, and all that can be done is to fence these off to prevent stock from picking up the parasites which infest these swampy zones. The drains themselves may need fencing to prevent stock from falling in or eroding the sides, and gates or stiles have to be provided where the public have right of access. General fencing of the area is also necessary to allow controlled grazing after improvement, and where it is impossible to fence – because the land is held in common or by a landlord hostile to enclosure – improvement is also impossible.

Herbage improvement can be achieved by renovation or by ploughing and reseeding, the choice depending upon the productive potential of the land and the cost of the exercise. If the land is wet and cannot be economically drained, surface improvement is better because ploughing tends to increase the rush content of the grazings. If the land is dry or can be successfully drained, most farmers prefer to plough because although it is more laborious and expensive, the boost in productivity is immediate. A persistent problem in the past was bracken, which is not killed by physical cultivation, but chemicals now exist which eliminate this invasive and poisonous plant.

Renovation of existing grazings involves the destruction of the surface mat of vegetation by some form of cultivation or by a quick burn, followed by the application of lime and fertilizer before oversowing with new grass. Reseeding involves the use of heavy equipment to remove any trees or bushes, and the land is ploughed to a depth of 20–30 cm. Heavy dressings of lime, phosphates and potash are applied and worked into the soil before the land is sown to a pioneer crop such as rape and Italian ryegrass. Sheep are allowed to graze this off in August–September, and the following year the surface is recultivated and reseeded with a permanent grass seeds mixture containing some rape, which helps the grass to establish. In some cases no pioneer crop is sown, but the land is immediately sown with the permanent grass mixture.

Improved utilization is the top priority once grazings have been improved. In the first few years, a high stocking rate for a short period is necessary if crops are to be eaten off quickly without poaching the soil and destroying the new sward. Once a tight sward has formed the field comes

Fig. 9.7 Land drainage prior to hill reclamation

Fig. 9.8 A very clear example of improved pasture on a bleak Welsh mountainside

into the normal rotational grazing system practised on the marginal farm where there is a high proportion of improved land, and improvement is continually being undertaken. On the mountain farm where unimproved land predominates, what little improved land there is is used only at critical periods when ewe nutrition has to be sustained at a high level (before mating and immediately after lambing). Though the proportion of improved land is thus smaller, it is arguably even more important on the true mountain farm than it is on the marginal farm, but it is generally improvement on the highest hills which arouses most criticism from conservationists.

The improvement of grazings entails higher stock numbers, which entail higher capital investment, so improvement should not be undertaken if capital is not available to exploit it. The extra stock also increases the labour and management demands, and the upkeep of the improved grazings further adds to the workload, for this is not a once-and-for-all job. Improved grazings rapidly revert to their natural state if not maintained, and bracken, heather and rushes quickly overwhelm the sown grasses. There is thus a constant need for applications of lime and fertilizer, for drains to be kept open and weeds controlled. In the process the natural habitat of some bird and animal species is destroyed, and for this reason as well as the loss of 'visual amenity' which results from the improvement of wild open hills, farmers are under increasing pressure to refrain from improving upland grazings. Some Exmoor farmers have been paid by local authorities to leave the moor unimproved, and it is likely that this initiative will be followed elsewhere as it is recognized that farmers ought to be compensated for foregoing income in the interests of conservation. Under the first agreement in 1978 one farmer undertook to leave 61 ha unimproved in return for a payment of £3,000 per annum for the period 1982–97, in lieu of lost income. Another farmer who wished to improve 174 ha based his claim for compensation on the following calculation:

Expected gross margin after improvement   £175 per ha/annum
Estimated costs of improvement   £125 per ha/annum
Margin foregone   £ 50 per ha/annum

On this basis the local authority agreed to pay him £40 per ha/annum, which he accepted although this entailed a loss of income of £10 per hectare.

SPECIAL TOPIC: FORESTRY IN THE UPLANDS

One of the problems of upland farming in Britain is that the hills are used for many purposes besides farming. Some are reserved for sport, many more for recreation, and a substantial proportion is used for both purposes in combination with commercial forestry. Though Britain was once covered with forests, today it has a lower proportion of its land area devoted to trees than any other European country (8 per cent, compared with 19 per cent in West Germany and 25 per cent in France). In 1980 in the UK there were about 2 million ha of forest and woodland, 75 per cent of which were managed for timber production. This total was composed of 1.36 million ha of conifers, which provide softwoods mainly for paper and board manufacture, though a small amount is used by the construction industry, and some 300,000 ha of hardwoods (oak, beech, elm) used for furniture making. Whereas the demand for hardwoods has declined steadily for many years, the demand for softwoods has constantly increased and is expected to double by the year 2025. The country imports 92 per cent of the timber and forestry products it uses, at a cost of £2,380 million, and throughout this century efforts have been made to increase domestic supplies by expanding the area of forest. In 1918 the Forestry Commission was established to ensure the continuity of supplies of home-grown timber (defined in the early days as reserves of 3–5 years), and by 1978 it owned 856,000 ha of forest. (Another 851,000 ha were in private ownership.) However, the Commission has never been able to acquire sufficient land to meet its obligations, which are now defined more widely as the provision of timber for industry, the protection and enhancement of the environment, and the provision of recreational facilities and support of rural economies.

For many years the Commission concentrated on growing conifers since they produce harvestable yields of timber in 40 years, compared with 100 years for hardwoods. The length of the production cycle is important to the economic viability of the enterprise, which depends on receiving a high enough price at the end of the growing period to cover the costs incurred and to provide an adequate return on the capital invested. In practice this means that the return must be higher than that derivable from alternative uses to which the land could be put, which depends on the profitability of alternative enterprises and on general interest rates. Forestry is therefore invariably concentrated in the uplands, for although trees will grow faster in lowland conditions, their maturing period is still far too long to make them competitive with agricultural enterprises. Since land prices in the uplands are also lower, reflecting the lower productive potential of the land, the concentration of forests in the hills is doubly inevitable.

The social advantage of this economic necessity is that since the uplands are sparsely populated, afforestation does not displace people from their homes. It is also beneficial in bringing employment to rural areas, since the labour requirement per hectare is higher in forestry than in efficiently structured agriculture. In addition it increases the provision of roads and other services which benefit the local population, and brings tourists into an economically weak region. In 1978 there were 24 million recreational visits to forests, and since tourists present opportunities to the rural population to provide accommodation and other services, this can bring more revenue to an area with comparatively few economic opportunities for the local community.

The possibility of multiple land use and ancillary economic benefits of this kind has encouraged the view that forestry should play a larger role in the economy. One proposal is that a further 2 million ha should be planted. This would increase self-sufficiency to 26 per cent by 2025, which would still only bring the forested area of the UK to 16 per cent. There are nevertheless disadvantages as well as advantages associated with afforestation. The uplands are often drinking-water catchment areas, and afforestation interferes with this, as well as reducing the amount of water available for hydro-electric schemes. The losses are unlikely to be large, but before major afforestation is carried out in catchment areas the costs of these losses must be assessed in relation to the extra revenue from timber. The same is true of sporting rights. In Scotland 270,000 red deer bring in revenue from the sale of sporting rights and meat, but the deer damage trees, and the high cost of fencing and the need to employ game-wardens counterbalance this income. Whereas it is possible to combine deer-keeping with forestry, however, grouse and forests simply will not mix, and in terms of economics forestry unquestionably wins. The comparable returns from the competing land uses of the uplands can be in the order of:

red deer	£ 1.40/ha
grouse and sheep	£ 9.77/ha
hill sheep	£24.50/ha
sheep and cattle	£92.00/ha
timber	£99.42/ha

These figures are simply indications, not national averages, and in many situations these are clearly not alternative enterprises. However, the orders of magnitude still suggest that in terms of income per hectare most sporting land should be afforested. The argument is not likely to persuade landowners dedicated to grouse shooting, but since it is possible to reconcile deer with forestry, economic necessity may tempt some of them to increase timber production on their land.

Most of the 3 million ha of land which it is technically feasible to plant with trees is currently farmed, and the Forestry Commission has never been able to acquire enough to meet its planting targets. Its aim is to purchase about 18,000 ha annually, but between 1970/71 and 1977/78 it averaged only 13,000 ha because of sustained opposition from the farming industry. Farmers inevitably consider food more important than timber, so they resist the conversion of livestock-rearing land into forest. Planting techniques in the past did not help the case for timber production. The Commission has often wanted to plant the middle slopes of hills (up to 450 m), causing serious problems for the farming system by creating a barrier which prevents the movement of stock between the in-bye and the hill top (Fig. 9.9). Small sheltered areas have also been planted up where sheep took refuge in severe weather, and the large blocks of forest make it difficult to burn rough grazings and heather to maintain their nutritional value. Large-scale afforestation has also been accused of isolating odd farms which become non-viable, so that their owners are forced to sell to the Commission. This fear has made it virtually impossible for the Commission to buy any land at all in some areas, especially where farmers are supported by conservation interests opposed to the afforestation of the hills.

Farmers' opposition to the expansion of timber production is understandable. They make their living from farming, and the irony is that as modern aids, subsidies and improved techniques are helping them to increase their incomes, the pressures on them to sell land for forestry are also

Fig. 9.9 Large plantings of conifers separate the in-bye from hilltop grazings

increasing. However, they have been so incensed by the threat which they identify in forestry that cases have been reported of farmers selling land to neighbours at lower prices than they could have achieved by selling to the Forestry Commission. The Ministry of Agriculture has generally backed their opposition, and since its approval is necessary before any land can be planted with trees, there has been an effective ban on the afforestation of the better upland grazings. A similar ban exists in some National Parks, where it is virtually impossible for anyone to plant trees without the Park Authorities' approval. In Snowdonia there is a *total* ban on tree planting.

The fact that forestry and agriculture compete for the same land does not inevitably entail conflict. Ideally they should be integrated into a multiple land use strategy which would combine the activities of forester and farmer and allow access to the public for recreation. Afforestation need not cause management problems for farmers if their needs are considered when the planting scheme is planned: forests can obviously provide shelter for stock and need not inhibit their movement. However, an increase in timber production must inevitably reduce the land area available for stock rearing, and in such a situation the livestock enterprise would have to be intensified, or stock numbers on the hills would decline. Unless there is a major re-orientation of agricultural policy, the former is more likely, since government support for hill farm incomes has always been based on the numbers of stock carried. Consequently, the upland landscape would be affected indirectly through the improvement of hill grazings as well as directly through the afforestation of larger areas of the hills.

Since both hill farming and forestry receive taxpayers' support, it is reasonable that the public should have a voice in deciding the use to which the uplands should be put. However, the public does not speak with one voice. Many people argue strongly for self-sufficiency in food, and would therefore prefer to see food production, not timber production, on the hills, importing timber from abroad. Others would argue equally strongly for greater self-sufficiency in timber products, which can be achieved only by reducing the land available for food production. Their case could be reinforced by the argument that agricultural output from the hills need not fall if the land were used more intensively through the improvement of grazings and increased stocking rates. Even if agricultural output did fall, this would be offset by increased timber production, since the value of timber imports is much higher than the likely increase in meat imports to substitute for the lost home production. The fall in agricultural incomes would also not necessarily mean that increased farm income support would be needed, since there would be improved economic opportunities for farmers from the forestry and recreational activities provided by multiple land use.

On the basis of such arguments the case has been put for much greater investment in forestry by the state and by private interests, and there are signs that traditional government support for hill farming may be revised to allow more support for the timber industry. To complicate the already complex economic arguments, however, there are aesthetic arguments about the 'loss of visual amenity' which results from the afforestation of the hills. The planting of large square blocks of conifers in the past by the Forestry Commission can certainly be criticized, but their recent planting schemes mix deciduous softwoods with evergreen conifers, and there are probably as many members of the public who believe that forests improve the upland landscape as there are those who deplore their presence. There is thus no easy aesthetic consensus, just as there is no easy economic agreement, so it almost certain that no coherent national hill land policy will be devised, because there are too many irreconcilable interests involved. What is likely is that some extra priority will be given to forestry without substantially reducing the aid to agriculture, and that wherever any proposal is made to alter hill land use (be it for agriculture, forestry, water conservation or recreation), the outcome will be the result of local disputes and compromises between the special interest groups concerned.

10

Mixed farming systems

Mixed farming systems involve the combination of arable and livestock enterprises on the same farm. The Norfolk Four Course was of course a mixed system, combining as it did wheat, barley and sheep production, and although there are all-arable or all-livestock systems in Britain, the majority of farmers still practise some form of mixed farming.

The techniques of crop and livestock husbandry are essentially similar to those used under single-enterprise conditions, but the need to combine them poses certain problems since the interactions are complex, and not fully understood. However, the combination of enterprises also provides opportunities to improve upon techniques or reduce the problems encountered on single-enterprise farms. These beneficial effects are stressed by the so-called *organic farmers*, whose motivations and methods are considered in the topic associated with this chapter. Several livestock enterprises are found almost exclusively on mixed farms, chiefly lowland sheep and beef finishing, because they rely on the by-products of arable enterprises as inputs or graze the leys which form an integral part of any mixed farming system. Pigs are also commonly associated with arable or mixed farms, but their interaction with the rest of the farm system is limited, so pig production will be considered later as a separate enterprise (Ch. 11). Milk production is also often found in association with cereal growing on many mixed farms, but again the production techniques do not differ from those used on specialist dairy units. On many mixed farms the dairy unit is in fact treated as if it were totally separate, with its own labour force which helps with other farm work only when it has nothing else to do.

There are many reasons governing a farmer's choice of a mixed farming system and the relative proportions of the different enterprises it combines. Some are related to climatic or economic considerations, or to the topography or layout of the farm. On a farm which consists of two blocks of land separated by land under other ownership, for instance, it might be necessary to have a dairy unit on one block and arable crops or beef or sheep on the other. However, one of the main reasons why farmers prefer mixed farming is the belief that it reduces the risks associated with a single-enterprise system – risk of crop failure resulting from bad weather or pest and disease damage, or risk of a price collapse in the one commodity produced, both of which would seriously reduce farm income. The risk of *all* the crop or livestock enterprises on a mixed farm suffering in the same year is by contrast remote, so such a system should not suffer from such wide income fluctuations, and in addition there are non-financial benefits like the ability to control weeds, pests and diseases by rotation.

Conversely, there are obviously problems associated with mixed farming. On a small farm it may be difficult to operate several enterprises, and it will certainly be more expensive if no one enterprise is on a scale large enough to achieve economies of size in the use of specialized machinery or buildings (though the use of contractors may make small hectarages feasible). Mixed farming may also entail a reduction in the possible farm revenue since it includes some less profitable enterprises, but this may be judged preferable to the losses that may arise on a single-enterprise farm. Mixed farming systems are moreover difficult to plan, and their day-to-day management is complicated. Where many different enterprises are combined, planning techniques like those discussed in Chapter 5 – especially linear programming – are invariably necessary, since it is essential to consider the interactions between the enterprises (e.g. the effect of farmyard manure from a beef or dairy herd on the yield and cost of potato growing, or the effect of a ley on soil texture and fertility). Another critical point is the design of a system which makes the optimum use of available labour. Many arable crops have a similar labour profile, a great deal of labour being needed in spring for sowing, and in autumn for harvesting. At other times of the year, however, little work is needed on the crops. On a livestock farm the labour requirement is more evenly spread. Labour is necessary in spring and early summer for lambing, calving and grass conservation, but it is also needed throughout the winter for feeding and early lambing. A combination of arable and livestock enterprises therefore provides employment throughout the year, which was traditionally one of the major attractions of a mixed farming system, and remains so on many farms today.

A typical labour profile of a mixed farm is shown in Fig. 10.1. There is a fairly constant demand for labour for the dairy enterprise and followers, and severe peaks in spring and autumn when the arable enterprises need attention and lambing and calving increase the workload. On this farm 1,200 manhours per month are available, which will not cover the spring and autumn peaks. Since the traditional employment of casual labour is not now feasible on most farms, the use of contractors for the spring and autumn cultivations may be the only solution. The cost or inconvenience of this might nevertheless persuade the farmer to modify the system to avoid the need for extra labour. The winter trough certainly means that the available labour is under-employed, but on a livestock farm there is always work to be done laying hedges, repairing fences, clearing ditches, and so on.

The two major enterprises of mixed farms not discussed in earlier chapters are lowland sheep and beef production, which are the backbone of the meat industry. As in the case of milk, the demand for meat is more or less

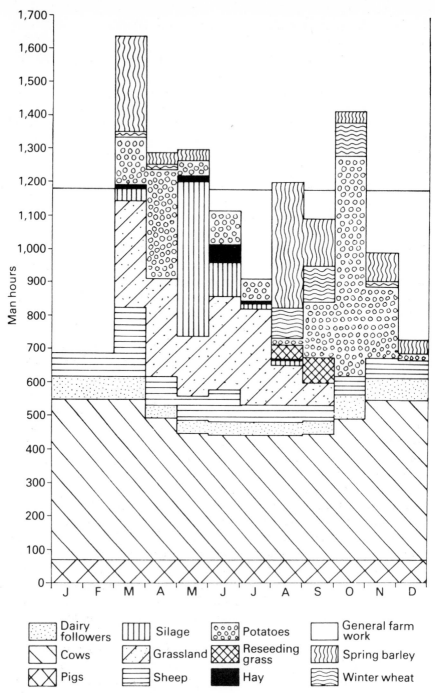

Fig. 10.1 A labour profile of a typical mixed livestock and arable farm (Longman Group Ltd. from *The Farm Business* L. Norman and R. B. Coote, 1971)

constant throughout the year, but production varies with the season. Most consumers are aware of the seasonality of lamb supplies, the result of a lambing pattern which is difficult to modify successfully. The first home-produced lamb reaches the market in late March and supplies increase to a peak in October. By mid-January domestic supplies virtually cease, and the gap until the new season's lamb arrives in the shops is filled by New Zealand imports. The seasonality of beef supplies is less well-known because most consumers are less aware of the origin of the beef they buy, but a similar pattern exists. The determining factor in this case is not the reproductive cycle of cattle (since, as Chapter 8 noted, cows can be made to conceive at any time), but the calving dates of the herds which produce calves for finishing on lowland farms. The main calving dates, we saw, are spring and early autumn, and the beef production pattern follows these peaks, with imported beef again meeting the intervening shortfall.

LOWLAND SHEEP PRODUCTION

On a farm which requires grass leys for rotational reasons, sheep have always had much to commend them. They are relatively cheap to buy and have a low capital requirement. Since they are lightweight they cause less damage to pasture than cattle, so they can utilize grass when cattle have to be taken off it. There are two main problems associated with sheep, however. They are notoriously difficult to constrain, so sheep farms have to be well fenced. On many farms where leys grazed by sheep would fit well into the farming system they therefore cannot be introduced because extra costs and labour for fencing would be necessary (though the use of lightweight electrified netting has reduced this problem). The other more serious problem is that they need shepherding. Someone on the farm must be able to cope with lambing and the complicated procedures involved in keeping sheep healthy and productive. On many arable farms such skilled labour simply cannot be provided, so sheep cannot be kept. (The labour required for shearing is not an obstacle, since contract shearers can be employed, as they are on many specialist sheep farms.)

Where sheep are integrated into a mixed farming system, it may be as *scavengers*, or as a permanent flock of breeding ewes.

Scavengers

Sheep on lowland farms are often bought from hill and mountain farms to be fattened on arable by-products, the aftermath of herbage seed production, sugar beet tops, waste from brussels sprouts or other vegetable crops and cereals. They thus literally act as scavengers, reducing waste to a minimum and providing an extra source of revenue at relatively little cost.

Sheep finishing varies widely from farm to farm. In some cases the lambs are kept for only a few weeks (where less than 5 kg liveweight gain is required), but elsewhere they may be kept until they are 10–12 months old (when a weight gain of about 12 kg is required). In the former case the farmer can rely almost entirely on by-products to feed the sheep, but when they are kept for longer periods special crops of swedes, turnips and rape are necessary to supplement the by-products. Even then, if a high stocking rate is carried, some concentrates may also be required. If such special crops are grown and concentrates fed, high stocking rates are essential to ensure an adequate return on the investment. (Typical rates would be 50–60 lambs/ha for a finishing period of 50 days on an average rape crop, or 100 lambs/ha for 10 weeks on an average swede crop.) One of the most popular scavenger systems involves the purchase of lambs and ewes as couples from mountain farms in spring (Ch. 9). All the problems of winter feeding and lambing are handled by a specialist stock rearer, and the lowland farmer simply fattens the lambs on grass for sale in June, July or August, and sells the ewes as drafts in August, or fattens them too for slaughter.

The economics of scavenger systems depends on the relationship between the price of store stock and the expected return from the finished sheep when sold for slaughter. To some extent the farmer can calculate the price he can afford to pay for sheep by deducting the cost of feed and other inputs from the seasonally adjusted EEC sheepmeat guide price (Fig. 10.3). However, when he grows special crops to feed scavengers these must be sown before he knows what the availability of sheep and the price of store stock are likely to be for the coming season. The small risk involved may nevertheless be considered worthwhile, bearing in mind the other non-financial benefits derived from sheep in the arable rotation. Scavenger flocks are therefore quite common, and many farmers plan to include them in a mixed system on a regular basis, though the enterprise may also be an

Fig. 10.2 A mixed flock of sheep feeding on a crop of swedes

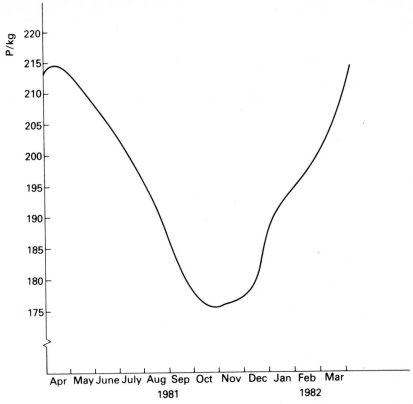

P/kg

Fig. 10.3 The EEC sheepmeat guide price for the marketing year 1981/2. This offers farmers an indication of the level of return, and therefore the price they can afford for bought-in stock.

opportunist one like multiple calf suckling on hill farms, depending upon the price at which store stock can be purchased in a particular year. The system is nonetheless declining as more hill farms fatten their own lambs, so where sheep are really valued in a lowland system, a permanent breeding flock may be introduced instead of scavengers.

Permanent breeding flock

Some lowland breeding flocks are purebred, kept for the production of breeding stock for sale to other farmers. A purebred Suffolk flock may be kept, for instance, for the production of rams sold for cross-breeding to fat-lamb producers. These flocks are self-contained, only the rams being bought in to prevent in-breeding problems. More common, however, is a system designed specifically to produce sheepmeat, based either on a flock of draft hill ewes or half-bred ewes for which replacements are regularly purchased.

Whichever system is practised, it must be run in such a way as to yield as high an income as possible compared with the other farm enterprises. In purely financial terms, we saw (Table 5.5), a sheep flock on a lowland farm can rarely be as profitable as arable cropping, but although they recognize this farmers still try to ensure as high a gross margin as possible. In doing so, lowland sheep producers have many advantages over their counterparts in the uplands. Since the climatic conditions are less severe hardiness is less important, so larger and more prolific sheep can be kept. The stocking rate can be higher because grass production per hectare is greater, again partly because of the better climate and longer growing season, but also because it is worthwhile to apply fertilizer. The stocking rate may also be increased without exacerbating the problem of worms because the cropping rotation allows the provision of clean grass for the sheep every year. There is, finally, more opportunity to improve the nutritional level by feeding arable by-products and specially grown fodder crops.

The profitability of the permanent breeding flock depends firstly on a high lambing percentage, and subsequently on the provision of a high level of nutrition so that lambs can be sold before the price is depressed by the arrival in the market of large numbers of upland lambs. As in the case of dairy cows, a compromise has to be made between the performance of the individual animal and the production per unit of land area, but (again as with the dairy herd) stocking rate is the most important element in the profitability of all grazing systems. In the case of sheep it is perfectly feasible to have a rate of 17 ewes plus their lambs per hectare and still produce satisfactory lamb carcases, provided reasonable nitrogen dressings (about 200 kg/ha) are applied. The need to sell lambs during the high-price period materially affects the time at which the ewes should lamb. Since the farmer knows from experience how long on average it takes for lambs to fatten on his farm, he can easily calculate the optimum lambing time. However, on a farm with arable crops which already has a spring labour peak, the lambing date may have to be adjusted to prevent it from coinciding with other spring work. In order to ensure efficient grass utilization the lambs should also ideally be old enough to make optimum use of the flush of grass growth which occurs on most lowland farms in May. In practice this means that most lowland flocks lamb between late January and early March, compared with late March to May on upland farms. Once the lambing date is determined, the entire management pattern of the flock is established for the year.

The breed of sheep which is kept is influenced by many factors, including the farmer's own preferences. There are fairly clear consumer preferences in respect of weight of lamb required and the degree of fatness of the carcase, and the different breeds vary with regard to these characteristics. The market requirement is for a lamb with a carcase weight between 15 and 22 kg, with a moderate fat cover (see Ch. 12). In choosing a breed to meet this specification farmers use the well-established principle that the optimum liveweight at which lambs on a high level of feeding should be slaughtered is half the weight of the mature animal of the breed. If this rule is followed, a carcase with the correct fat cover, weighing a quarter of the mature weight will result. Most fat lambs are the result of crosses with a domestic breed like the Suffolk or an imported breed like the Texel, which produce carcases with a high lean-to-fat and lean-to-bone ratio which are ideal for meat production. The optimum slaughter weight can thus be determined given information on the mature breed weights. For example, a Suffolk × Welsh Mountain lamb should be slaughtered at 34 kg liveweight. If the farmer wanted a larger lamb, but still preferred to use a Suffolk ram, he would have to use a bigger ewe, say a Welsh Half-bred. Alternatively he might retain the ewes and use a larger breed of ram like the Oxford Down. What he must *not* do is to keep his Suffolk × Welsh Mountain lambs longer and increase their weight, since the extra weight is put on as fat, producing a carcase that will be down-graded.

Prolificacy is another important factor determining the choice of breed, since it obviously affects the actual number of lambs available for sale. There are distinct breed differences in prolificacy, and the cross-bred ewes (e.g. Masham, Scotch/Welsh Half-bred, Mule) used for further crossing with a fat-lamb sire generally have higher lambing percentages than the hill ewes from which they originated. They are therefore ideally suited for the

Fig. 10.4 A pedigree Suffolk ram used to produce a high-quality meat carcase. The ram's harness holds the coloured crayon which marks ewes at mating

better conditions of lowland farms, and since the lowland farmer can also keep his ewes in better condition and provide a higher level of nutrition than the upland farmer, they can also rear more than one lamb each.

The condition of the ewe at mating has a significant effect on ovulation and the number of lambs born. A major part of ewe management is therefore to ensure that the ewes' condition is improving at mating time, which means improving the level of nutrition 3–4 weeks before mating. This is again easier to achieve in lowland conditions than it is on the hills. Unlike cattle, which will breed at any time of the year, most breeds of ewe only come into oestrus in a period of declining day length. Mating therefore takes place from late August onwards, the exact time being fixed by the farmer so that lambs are born when it best suits his farming schedule. The gestation period is approximately 147 days, so if lambs are to be born in February, the rams are run with the ewes in September. The rams normally have a coloured crayon secured to their breast bone which marks the ewes at mating, the colour being changed every 14 days. This enables the farmer to separate the flock into groups according to their expected lambing date, which allows for better management. After mating the ewes are set-stocked at a low stocking rate, but approximately six weeks before lambing they receive some concentrates plus hay or silage. During these last six weeks the foetus makes two thirds of its growth, and since this period coincides with the greatest shortage of grazing, a high protein concentrate feed is provided. This is introduced gradually, but immediately before lambing about 0.35–0.65 kg/day should be fed, according to the size of the ewe. Some farmers also feed root crops at this time, but they are not a complete substitute for concentrates because the ewe cannot eat enough of them to provide the energy required.

On many farms ewes are housed as the lambing date approaches, at least during the night. Special buildings are not necessary for this prupose, since general-purpose buildings which serve as hay, potato or other stores will generally be empty by this time. Where ewes are not housed, at least some shelter is normally provided in which they can lamb, and after lambing the couples are turned into a sheltered field which has not had sheep in it. Since the lambs are dependent on the ewes' milk, the ewes' nutrition must be good, so concentrate feeding generally continues for several weeks, especially in flocks which lamb early in the year. Once the grass is growing rapidly the ewes and lambs are set-stocked. During the grazing period the lambs may gain weight at the rate of 250 g/day, so a lamb with a birth weight of 4 kg would reach 34 kg in four months. With a February lambing, a good sheep farmer could therefore expect to be selling lambs in June and July. These lambs are sold directly off the ewes, but many farmers wean lambs after 12–15 weeks as grass growth declines, so that the ewes can graze poor pasture while the lambs are given the best young grass, reaching slaughter weight in about 16 weeks. The stocking rate during this period on most

Fig. 10.5 Ewes and lambs feeding in winter housing (Farmers Weekly)

farms is not much more than 10 ewes plus lambs per hectare, though it is possible to raise this to 12 by good management. It can be raised even further if rotational paddock grazing is practised, but this is uncommon because of the high fixed costs involved in fencing. Of course, the stocking rate declines as lambs are sold off throughout the summer, so that by mating time there may be only the ewes left on the farm.

The worst management problems of sheep are the innumerable diseases, worms and parasites from which they suffer, and constant surveillance and treatment are necessary if losses are to be reduced. This is even more important since the sheep enterprise is already comparatively unprofitable under lowland conditions. One major limitation on the profitability of sheep is the fact that ewes lamb only once a year, and even then only produce two, or very occasionally three, lambs. The answer is not to seek a higher lambing percentage, since if a ewe produces more than two lambs artificial rearing is necessary, and this has restricted the use of the few breeds (like the Finnish Landrace) which do produce litters of up to five lambs. The other way of increasing lamb production which has been tried is to take two lamb crops per year, which should be possible since the gestation period is only five months. It has long been known that one British breed, the Dorset Horn, will lamb out of season, and this has been capitalized on by producers who specialize in providing lamb for the luxury Easter market. These producers lamb their ewes in November and December, and use concentrates to pro-

duce a lamb of 40 kg weight in ninety days. Considerable effort and research have been devoted to this and other breeds in the attempt to obtain
two lamb crops a year, but the success has been limited – though there are
some flocks which produce three crops in two years. The system is nevertheless complicated, and involves the early weaning of lambs (within eight
weeks of birth) and their subsequent artificial rearing. Hormones and controlled day length patterns are also used to stimulate ovulation in the ewes.
One experimental flock of Finnish Landrace × Dorset Horn ewes has produced an average of 3.3 lambs per ewe per year, but the management problems are too great to show much promise for the widespread adoption of the
system on commercial farms in the near future.

BEEF FINISHING

Beef production is the second major meat production system found on
mixed lowland farms, its chief input being weaned calves purchased from
upland farms or surplus calves from the dairy herd. The degree of integration of dairy and beef production is a phenomenon which is almost unique
to the UK, and is certainly not found in the USA, Australia or South America. Since the dairy herd produces far more calves than it needs for replacements, most cows are mated with a beef bull to produce cross-bred heifers
(many of which become the upland farmer's suckler cows) and cross-bred
bull calves which produce 58 per cent of the beef raised in Britain.

There are probably more methods of finishing beef cattle than there are
of producing any other agricultural product. This results from the fact that
the enterprise is always a secondary one, even on mixed farms, and therefore has to be integrated into the rest of the farming system. The type of
stock used also varies very considerably, as does the feeding system. There
are nevertheless certain fairly well-established planned finishing methods for
cross-bred dairy and weaned calves which this section will outline. Again,
however, there are also many opportunist systems of beef finishing practised
generally by older farmers who relied on imported Irish store cattle which
they bought when it was cheap and they had some spare grazing on the
farm. Once bought, these cattle were grazed in a very traditional way, with
generally high standards of management which produced a daily liveweight
gain of up to 1.5 kg with a stocking rate of three animals per hectare in
May. At this rate, some of the cattle could be sold off in June, so that the
stocking rate fell, making more grazing available for the remaining stock.

This opportunistic behaviour is now less common than it was, partly because the Irish store trade on which it was based has declined sharply, but it
still exists in specialist beef-producing areas, and it demands experience and
luck to make money out of it. For example, in February 1978 beef finishers
were buying store cattle at 67.9 p/kg, and a typical budget for the finishing
process was:

	£
280 kg store at 67.9 pence	190
Summer feed	15
Winter feed to December	50
Other costs	15
Total variable costs	270
Fixed costs	50
Reasonable profit	40
	360

To break even, in other words, the finisher needed to obtain £320 per finished store, and £360 if he expected a reasonable return on the investment in capital and labour. If the animal weighed 470 kg, the sale price would have to be 68 p/kg for break-even, or 76.6 p/kg for profit. In fact the best estimate at the time of purchase of the likely realization price 10 months later was 67.5 p/kg, and the actual price they fetched was 66.7. This suggests that finishers were prepared to break even rather than make a profit, and that they were paying too much for store cattle, basing the price they paid not on likely *future* beef prices but on the price of beef in February, when it was high because supplies were short.

The object of beef production is to produce a carcase of a weight, shape and yield of saleable meat required by the market, in a manner which is profitable for the farmer. This is not easy, if for no other reason than that there is conflict between the price and eating quality required by consumers and the butcher's desire for a high yield of saleable meat, a high proportion of it in the form of high-priced cuts. There is further conflict between the beef finisher, who is only interested in transforming feed into beef, and the calf producer, who is interested in the performance of both calf *and* cow. This conflict is most readily seen in the case of the upland farmer who may have to keep a given breed for its hardiness, although he knows that the finishing performance of the weaned calf may not be very high. On the dairy farm the problem is less acute, since the Friesian breed produces crosses which are fairly well-adapted for beef finishing. There are problems, however, with more specialized dairy breeds like the Guernsey or Holstein, whose calves do not fatten very well.

Beef from dairy calves

The finishing of cross-bred dairy calves has been extensively studied in the attempt to devise a method which will allow the farmer to produce good quality beef at a reasonable profit. The most popular system was developed at the Grassland Research Institute for 18-month old beef from autumn-born calves. The system is based on the use of high-quality forage and conserved products and concentrates, but the amounts of each used can be very flexible, as may be the type of calf used. The essence of all finishing methods, however, is identical: to adjust the amount of feed as accurately as possible to the stage of growth and the breed of cattle, so that feed costs are kept to a minimum, while simultaneously ensuring that a satisfactory carcase is produced.

During the growth and development of an animal the ratio of muscle (= meat, which we require) to bone and fat (which we do not require) changes. After birth the muscle-to-bone ratio increases as the animal grows to maturity, after which most of the energy in its feed begins to be transformed into fat. The rate of growth varies from breed to breed. Many British breeds (e.g. Hereford, Aberdeen Angus) mature early, while European breeds like Charolais and Simmental are late-maturing. There is therefore a tendency for British breeds to put on fat while the European breeds are very lean at slaughter weight, which is why they have been introduced into Britain in recent years. It is nevertheless possible to adjust the feeding of early-maturing breeds to produce meat without fat. The same diet fed to a late-maturing breed would limit the production of muscle as well as fat, so farmers clearly need to make the feeding regime fit the breed very closely. However, the evidence is that many farmers do not adjust the feeding regime sufficiently to suit the breed they keep, and the resulting high proportion of carcases which do not suit consumer requirements led to the introduction of recom-

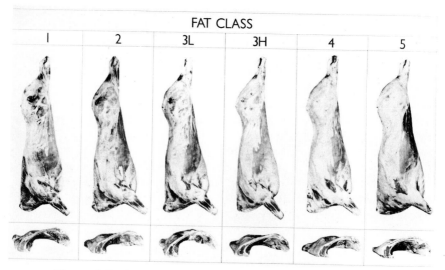

FAT CLASS

| I | 2 | 3L | 3H | 4 | 5 |

Fig. 10.6 Beef carcases classified by fatness. The preferred classes for most of the trade are 3L and 3H, but there is evidence that consumers prefer Class 2, which is leaner (M.L.C.)

mended carcase specifications by the Meat and Livestock Commission, whose long-term object is to bring beef production more closely into line with the market requirements.

Finishing potential varies according to the sex of the animal as well as its breed. Bulls grow faster and produce a leaner carcase than steers (castrated males), and steers are leaner than heifers. (The main reason for castrating bulls is not a farming one at all, but the requirement to reduce the danger to the public which has access to fields where they graze.) In deciding upon his system of finishing the farmer must take all these factors into account, and in practice he purchases groups of cattle of the same breed and same sex, with a similar weight when they are bought, and this simplifies the feeding regime within the herd. Having decided the optimum mature weight to aim for the feeding regime can be determined, perhaps in accordance with a system like that developed by the Grassland Research Institute for Hereford × Friesian steers, summarized in Fig. 10.7.

This cross is able to make good use of cheaper forage products, and there is a ready source of the basic input from the dairy herd. Autumn-born calves are bought from the dairy farmer when they are between 1 to 3 weeks old. Many are bought through auction markets, but there are risks of disease, and it may be difficult to acquire enough evenly-matched calves at auction. One solution has been the formation of weaned-calf cooperative marketing groups which assemble calves from their dairy-farming members and sell them direct to finishers. This eliminates the disease risk of the auction mart, and the cooperative can match calves evenly to suit buyers' requirements.

During the winter months the calves are fed rations sufficient to gain weight at 0.8 kg/day, so that they weigh approximately 180 kg when turned out to pasture in the spring. The feed does not contain a high proportion of concentrates, because after a period of relatively poor diet they put on weight quickly when turned out to fresh grass in the spring. This saves feeding costs during the winter, yet produces a calf by the end of the sum-

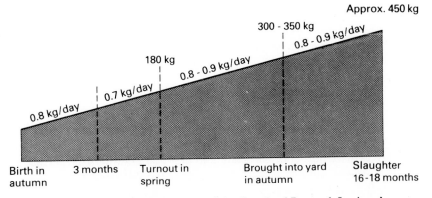

Fig. 10.7 A simplified version of the Grassland Research Institute's recommended system for beef production from Hereford × Friesian steers.

mer which is indistinguishable from one which was well-fed throughout its first winter.

The calves are stocked on clean pasture at a rate of 8.5/ha during the period of peak grass growth, reduced to 5.0/ha as the summer progresses. The optimum rate of liveweight gain during this period is 0.8–0.9 kg/day. To achieve these high stocking rates without inviting severe parasitization it is usual to paddock graze the calves, though carefully managed set-stocking can produce equally good results. This regime produces cattle weighing between 300–350 kg when they are brought indoors for their second winter, which is the real finishing period. The rate of gain in this period is still only about 0.8 kg/day if a slaughter weight of approximately 450 kg is to be achieved (the ideal weight for the Hereford × Friesian). This needs a daily ration of some 26 kg of high-quality silage and 0.75 kg of rolled barley plus added minerals and vitamins. The overall stocking rate for the system for grazing and conservation will thus be approximately 4 cattle/ha, and during the finishing period of the second winter (165 days) the cattle will consume between 3 and 5 tonnes of silage and 125–300 kg of barley, depending on the quality of the silage. High-quality silage is very important for the economics of this system, since poor-quality silage needs supplementing with up to three times as much barley. The crucial importance of good grassland management to the economic viability of beef production is thus obvious, and it is even more crucial here than it is in dairy farming because of the comparatively low gross margin of the enterprise. It is for this reason that most beef finishers now feed silage rather than hay, because it is too risky to rely on the vagaries of the weather to produce a hay crop good enough not to need even higher supplements of barley to achieve slaughter weight.

For the same reason, there has been an increasing use of growth promoters by both beef and sheep finishers, which increase the liveweight gain and boost the conversion of feed into saleable meat. Hormone implants are the commonest method, which has received much adverse publicity in the press, raising fears that consumers of the final meat product may suffer hormonal imbalance from eating it. The evidence is in fact that there is no real danger of this occurring, and the use of these aids is strictly regulated, but there has nevertheless been sufficient disquiet to lead to proposals that legislation should be introduced to prevent their use altogether.

Beef from suckled calves

In this system the finisher does not rear calves at all, but purchases upland autumn-born calves when they weigh about 290 kg, or winter- and spring-born calves weighing approximately 230 kg. These he finishes during the winter at 15–18 months of age, or off grass the following summer at 24 months. The same considerations determine the system here, chiefly the need to have fairly well-matched groups of calves which facilitate the feeding regime. The finishing period is likely to be fairly short, about 160 days, and once the calves are purchased and feeding starts it is very difficult to react to market changes, so that the farmer's flexibility of management is limited. The key decision is therefore the price he can afford to pay for calves and still make a profit, which depends on the expected sale value of the finished cattle and the cost of feed and other inputs. The expected sale value and feed costs vary, of course, with the length of the finishing period and the type of cattle being fed. A good finisher can achieve a daily liveweight gain of 0.9 kg, so that in 160 days the cattle gain 140 kg. The fundamental ration will again be high-quality silage or hay, supplemented by up to 2.5 kg/day of concentrates, some 440 kg of concentrates being required, perhaps, for the whole finishing period. This regime is successful with autumn-born calves, but the smaller spring-born calves are normally kept as stores through their first winter, gaining about 0.5 kg/day, and finished on grass in their second summer.

Beef finishing under both systems is a difficult and risky enterprise since cattle often have to be kept for up to two years, during which the cost of feed and the price of finished cattle can change significantly, and interest rates on money borrowed to purchase cattle and feed can move alarmingly. It has frequently been argued, in fact, that only farmers with a canny *feel* for the market ever make any money out of beef finishing. Several recent studies have nevertheless shown that in most situations the difference between the best and worst financial returns is not the result of different market prices, but of differences in cattle performance and feed costs. Clearly a *feel* for the market may help, but the successful beef finisher today is not a cattle dealer with land, but a skilled stockman maximizing his margin through careful selection of stock, meticulous attention to feeding, and generally high management of animals and grass.

CONCLUSION

This chapter has concentrated on beef and sheep finishing as the two major enterprises of mixed farming systems which have not already been discussed, or will not be discussed in the following chapter which touches on pig and poultry production. Ideally it should now be possible to draw together the components of the 'average' mixed farming system, but given the almost unlimited permutations of enterprises combined on real mixed farms the exercise is artificial and not ultimately useful. Beef and sheep finishing will loom large in most systems, combined with arable crops which integrate well with them. What determines whether or not different enterprises fit well together will very often be the labour situation on the farm, and Fig. 10.1 illustrated very clearly the intricate labour interactions of a multi-product farming system. A similarly complex picture could be drawn up for the same farm in relation to capital requirements and revenue. The financial anchor of this farm is the dairy enterprise, bringing in a monthly milk cheque, supplemented by fairly regular revenue from sales of pigs. This regular income will probably suffice to sustain the costs of these two enterprises as well as subsidizing others which produce income at less regular intervals. Income from the arable crops and sheep, for instance, peaks at certain points of the year, as do capital requirements for seed, fertilizer, fuel, feed, etc., and careful financial as well as technical management is needed to maintain the farm's economic viability. The design of the system is thus more crucial on a mixed farm than in any other situation, and because the interactions are so much more complex, even a slight adjustment in one component can set off a train of reactions throughout the rest of the system.

For this reason, it is probably the case that the best *mixed* farmers are the best *farmers*, which is a paradox that underlines the major difference between agriculture and other production processes where specialization is generally the key to excellence. By extension, it should be true that the best farms are mixed farms, and this is certainly the argument of the so-called organic farmers and a growing number of *low-input/low-output* conventional farmers whose system is examined in the concluding special topic.

SPECIAL TOPIC: ORGANIC FARMING

Organic farming is popularly associated with a back-to-nature movement which rejects modern agricultural methods out of hand. In its less extreme form, however, *low-input farming* (as it is better described) is a serious and growing effort to reduce dependence on inorganic fertilizer and chemical controls without drastically reducing the industry's capacity to feed the world's growing population. Earlier chapters have emphasized the fact that economic objectives are frequently modified by technical constraints; that most farmers are less interested in maximizing short-term profits than in ensuring a satisfactory income for the whole of their working lives. The limited control achieved by agrochemicals has also been noted. The only universally effective long-term control for many pests and diseases, it was suggested, is the old principle of rotation and high management standards. For many farmers all this points to the wisdom of a mixed farming system, for although revenue is reduced by the inclusion of low-margin enterprises, financial risk and costs are also reduced, and the farm's long-term viability is maintained by the interaction of enterprises which sustain soil fertility and a healthy farm environment. Exactly the same reasoning informs the principles of organic farming, and in practice there is so little to distinguish the best mixed farms from an organic farm that some of them have had to make only slight modifications to their system in order to be able to market 'organically produced' food, which fetches a premium in a small specialist market. In fact, the only distinction between good conventional mixed farmers and organic farmers is that the latter are generally dedicated evangelists who set out actively to convert others to the cause.

Although there are organic or low-input farms which practise all-livestock systems, or a mainly arable system in which short leys are introduced only as true break crops, most organic farms are mixed enterprise systems. Indeed, one of the principles of organic farming identified by a Soil Association pamphlet (Mayall, 1976, see Further Reading) is 'practising mixed husbandry within the limits imposed by physical conditions and economics'. The object is to make 'as much use as possible of natural resources, . . . avoiding the use of chemicals liable to kill or reduce the activity of soil organisms . . . and avoiding direct feeding with soluble minerals'. The vocabulary is revealing. The object is not to proscribe altogether, but to *avoid* the use of artificial fertilizers and chemical controls, and research into commercial organic farms in the UK has suggested that in practice they are *not* prohibited. The best organic farming, like the best conventional modern farming, is thus prepared to compromise when it is necessary, and only purists treat it as an end in itself rather than a prudent means of producing income and food 'within the limits imposed by physical conditions and economics'.

In order to avoid as far as possible the use of inorganic fertilizers, the organic farmer and the low-input mixed farmer return to the soil all animal

Fig. 10.8 Organic farming rejects the high artificial fertilizer dependence of modern agriculture (Farmers Weekly)

and vegetable residues, and seek to maintain a healthy population of soil micro-organisms to transform the recycled waste into usable plant food and humus. In addition they use organic manures (including compost, seaweed and fish manures) and some 'permitted' fertilizers like basic slag (a by-product of an industrial process) and rock phosphate. This policy is combined with one of minimal soil disturbance by cultivation, the object of which is to develop an open soil structure which makes air and moisture available and sustains a high enough temperature to encourage the conversion of wastes into usable plant nutrients. It is therefore imperative not to plough deeply or cultivate the topsoil so much that this structure is destroyed.

The control of weeds, pest and diseases on organic farms is achieved as far as possible by means of the husbandry practices which are used on conventional farms in combination with agrochemicals, the object being to reinforce the natural or inherited immunity of plants and animals. It is claimed that rotational cropping, physical hoeing and undersowing of crops with clover suppress most weeds, but herbicides which degrade rapidly in the soil are used where these measures prove insufficient. No synthetic insecticides or fungicides are permissible, however, because they persist in the soil. On a well-run organic farm, it is argued, fungus diseases simply should not occur, and pests should be kept in check by their natural predators. One serious problem for which no organic control has been found, however, is potato blight, so all but the real purists will use chemical sprays in this case. Some organic insecticides are also used (e.g. pyrethrum and derris) because they do not persist in the soil and are non-toxic to humans.

Livestock on organic farms in the UK are kept in very similar conditions to those kept on conventional farms, the main difference being that the stock are generally mixed, and fed as far as possible on home-produced crops. The other main difference relates to the use of antibiotics and other drugs, which some conventional farms use routinely as a precaution against disease and parasitization, while organic farmers use them only when natural resistance is overcome by a severe outbreak. Needless to say, all artificial growth aids like hormone implants are also outlawed from organic systems. The contrast between livestock kept on organic and conventional farms is much greater in North America and Australia, where a high proportion of beef cattle and some sheep are kept in intensive feedlots like battery hens. Under these high-stocking conditions routine dosing is necessary because of the increased risk of disease, and the animals' diet is controlled down to the last gram, and heavily dependent on concentrates. In Britain such feedlots are still relatively rare, and only pigs and poultry are kept in intensive conditions (Ch. 11), which are of course rejected by organic farmers on grounds of good management as well as in the name of animal welfare.

In organic farming as in mixed farming, the critical function is the design of the total system – called in this case the farm ecosystem. Rotation is the foundation of the system, supported by high management standards and scrupulous attention to hygiene. From the economic point of view the objective is never maximum production. Organic farmers tend to believe that high yields mean lower quality, so quality rather than quantity is their guiding principle. This has recently appealed to a growing consumer demand for quality which has provided organic farmers with a remunerative specialist market. Organic food shops have opened throughout the developed west, but their prices are often twice those of local supermarkets, and as long as most consumers continue to value cheapness more than high quality the market will not expand rapidly. Advocates of organic farming argue that high retail prices for organic food result from the high costs involved in handling the small amounts of produce currently reaching the market from very dispersed sources, and that as in the case of most industrial products the price would fall if output increased. Sceptics reply that though this may be true, the production costs of organic agriculture are themselves higher, and since they would probably not fall with increasing output, retail prices would still reflect this higher production cost. One American estimate is that organic farming methods would add about 10 per cent to the production costs of the major field crops, but against this organic advocates maintain that since only about 38 per cent of retail food prices is accounted for by farm production costs, the increase in retail prices should be much less. If in addition the possible health advantages of eating wholesome food are considered, they maintain, and the savings in non-renewable energy resources used to manufacture inorganic fertilizers, the extra cost is a small price to pay. Finally, though they admit that organic agriculture cannot currently match the production levels of conventional farming methods, they argue that if the same amount of research effort were devoted to the improvement of organic farming that has gone into conventional agriculture in the last half century, its output could almost certainly be increased.

This would answer the serious objection to organic farming that it cannot currently feed the world's population. The argument for quality rather than quantity is a luxury which only a well-fed population can afford, and it is difficult to persuade developing countries that they do not need artificial fertilizers and chemical controls to raise their level of nutrition to something like the standard enjoyed in developed countries. In the last fifty years the use of fertilizers and agrochemicals increased grain yields in the USA alone by 250 per cent, making it the bread basket of many Third World and even some Communist bloc countries. How many years will it take, given the greater complexity of the biological science on which organic farming seeks to base its advance, to achieve similar results without chemical fertilizers or controls? One answer is of course that however long it takes, this is the only sound long-term policy which agriculture can follow. Fertilizers and chemicals are manufactured from non-renewable resources, in an industrial process with a high energy requirement, and their price is certain to continue rising as these resources are exhausted. Both fertilizers and chemicals also unquestionably have damaging effects on the natural environment, but the reality is still that people value food more than they value the environment, and that agriculture cannot abandon either without sharply cutting food production.

Equally, agriculture has never been static, and it will go on changing and developing substitutes for the inputs on which it is currently dependent. Dependence on chemical inputs as they now exist will probably decline, and the mixed farming systems which most farmers still prefer will probably prove even more attractive than they already do. Low-input/low-output farming may even in time conceivably supersede the high-input/high-output systems of the last twenty-five years, especially now that *governments* are showing interest in its ability to reduce agriculture's dependence on scarce energy sources. What should be clear, however, is that this will not represent a return to a farming tradition temporarily abandoned, but a new manifestation of the capacity for self-adjustment to a changing environment which agriculture has always displayed.

I I

Intensive livestock

Intensive livestock systems have very high stocking rates on a very limited land area. In Britain such systems are confined mainly to pig and poultry production, though there is some intensive veal calf production and some intensive beef. Land is in a sense only incidental to these systems, and it is perfectly possible to have a substantial intensive livestock business on only ¼ ha of concrete. Land is thus necessary mainly as a provider of feeding-stuffs.

It is in fact legitimate to ask whether such *factory farming* can be considered part of agriculture. Agricultural textbooks already exclude intensive poultry production on the grounds that it has become an industrial process, controlled by a handful of large producers with little or no knowledge of conventional agriculture. Pig *finishing* is also becoming a large-scale process with little in common with conventional farming systems, but the breeding and rearing of weaner pigs is still mainly a farm-based activity which contributes quite substantially to many farm incomes. Precisely because it required little land, pig production was always an ideal enterprise for the small family farm, but the existence of large-scale industrial units is making it a less reliable source of income for the small producer. Because an intensive system can be brought into production comparatively quickly, it tends to be a speculative enterprise which attracts people who respond to sudden increases in the market price. The result is periodic gluts and shortages of pigs as prices cycle from high to low and back to high, and the profitability of the enterprise is subject to swift changes. This tends to reinforce the concentration of production in fewer hands as small producers decide that the uncertainty and risk are too great to bear.

Pig and poultry production also lend themselves better than any other livestock system to the kind of control which characterizes industrial production lines. Every component in the system is directly controllable by man. The effect of weather, for instance, is eliminated by completely enclosed controlled environmental housing. Feed can be very carefully formulated and very efficiently fed. Even the day length and light intensity are generally controlled automatically, in order to regulate the rate of egg production for instance. It is possible to operate very strict disease control programmes, so that very little is left to chance, and the entire system regulated more closely than is imaginable in any other agricultural enterprise. The marketing is also more tightly organized, and the producer is often not a farmer at all, but a large food company or retail chain. The production of eggs and poultry meat are already often vertically integrated systems, and the same is increasingly true of pig production, which is often in the hands of retail food companies or feed manufacturers.

This also applies to a large extent to beef and some sheep production in North America and Australia. Feedlots exist where anything between 10,000 and 100,000 cattle are kept in intensive conditions like poultry and pigs in the UK. One argument for this is that livestock production would otherwise be difficult in a climate which supports insufficient and inferior herbage. Another reason is an alleged consumer preference for *maize* – (not grass) fed beef, but it is also true that the system attracts investors, and is admittedly more efficient than traditional extensive grazing systems. In Britain and Europe none of these motives have been very persuasive, and the only significant incentive to introduce intensive beef feedlots is the shortage and high price of land. The only intensive cattle systems in Europe are veal production units, which are extremely uncommon in Britain but more common on the Continent where the consumer demand for veal is greater. This chapter therefore concentrates mainly on pig production, on the grounds that it is the only significant intensive livestock system on British farms, whose general management principles may illustrate the broad principles applicable in other intensive systems. In the special topic the discussion is broadened to examine animal welfare, one of the most contentious aspects of intensive systems brought to public attention chiefly in relation to poultry and veal calf production.

PIG PRODUCTION

On some farms pig production is the principal or only source of income. Since it integrates well with many other farm enterprises, however, it is a subsidiary enterprise on many other farms where it makes a worthwhile addition to the farm revenue. Pigs fit particularly well into a mixed farming system where they consume grain produced on the farm, and their manure can be disposed of easily and to good effect on the arable land. The ultimate object of pig production is to convert vegetable feedingstuffs into meat for human consumption as pork, bacon, or one of very many processed meat products (sausages, luncheon meats, patés, pies, etc.). Pigs in fact produce 50 per cent of all meat consumed in the world, and their feed conversion efficiency is unrivalled by other animals. The economic viability of the system relies entirely on this feed conversion efficiency – that is, the amount of food necessary to produce a given quantity of saleable meat. This depends on the animals' breeding and management, both of which have been the object of intensive research. The importance of the feed conversion efficiency at the *finishing* stage has been especially studied, since feed at this stage represents 43 per cent of the total costs of pig production, and considerable advances have been made towards devising an optimum feeding regime.

As important as feed conversion efficiency is the reproductive efficiency of the sow, which is after all kept only to produce piglets for finishing. The skills required for the management of sows and for the finishing process are not identical. The former demands a very high degree of stockmanship, but finishing requires primarily an ability to organize a complex procurement and marketing operation, because its profitability depends on the complex interrelation of the price of pigmeat and the price of feedingstuffs. Finishing is in fact a problem of economic management rather than of stockmanship, which is why it has been possible to remove it from farms into an industrial environment. The different skills long ago led to a division of labour within pig production, however, into three separate but closely integrated sectors. The breeding sector provides the basic breeding stock; weaner producers rear pigs to approximately 8 weeks of age (18 kg), and they are then sold to finishers who feed the pigs to pork or bacon weight. The three sectors are integrated so that the finisher is certain of obtaining the right type of weaner for his particular system, since a pig bred for feeding to pork weight (70–80 kg) is unsuitable for bacon production (100–120 kg). Many contractual arrangements therefore exist to ensure that the weaner producer obtains the right type of breeding stock in order to produce the right input for the finishing system. The ultimate way to ensure this is, however, for the entire production process to be under the same ownership and control, and it is this that is eroding the traditional division of labour within the industry and leading to completely integrated units.

Weaner production

A weaner production system depends on stockmanship of a very high order, the object being to produce and rear as many piglets as possible to as high a condition as possible, which requires time and attention to detail. The welfare of the sow and her progeny are at this stage paramount, and considerations like slight savings in feed or labour costs should take second place because the improvements they make to the profitability of the system are also secondary.

The pig is one of the few mammals which normally produces multiple offspring. The average number born alive in UK herds is just over 10, and has remained static over the last decade. Litter size is related to the sow's condition, which depends on her feeding. The sow does not reach mature weight until she has had about five litters, so her nutritional level must be sufficient to sustain her own growth (ideally she should put on 10–15 kg between each pregnancy) as well as support the foetuses. The nutrition of *gilts* (sows pregnant for the first time) is particularly important, and as with sheep it is important to improve the level of feeding before mating, since this increases the ovulation rate and the potential size of the litter. The sow's reproductive processes cause many management problems. Not only does the gestation period vary in length (though the range for 92 per cent of sows is only from 113 to 119 days); the oestrus period, which only lasts about 60 hours, is also very difficult to detect. This makes it difficult to arrange mating at the right time, and if mating takes place too soon or too late the litter is reduced by as many as three to four pigs. Lacking any reliable means of detecting the optimum mating time, many pig keepers therefore mate the sow twice during oestrus, which significantly increases the litter size. A further complication is a degree of seasonality in the breeding cycle. In western Europe, for example, there seems to be a period of low fertility in sows weaned in June – September, the causes of which are unknown. Certain breeds also produce larger litters than others, and crossbred or hybrid sows generally produce more live pigs. All these factors can increase the performance of the herd, but more important is that throughout pregnancy the sow or gilt is comfortable and contented, which depends on adequate housing and nutrition.

Efforts to increase litter size are wasted if they are not backed up by high management standards to reduce post-natal losses. The aim should be a survival rate of at least 95 per cent, which the best-managed herds achieve regularly. Average losses after birth are much higher, however, though there was a slight fall from 15.2 per cent in 1970 to 13.6 per cent in 1979. Half of all losses occur in the first two days of life, and even later losses frequently result from incidents during this critical period. The commonest causes of post-natal death are starvation and crushing by the sow, which together account for between 50 per cent and 80 per cent of total losses, though both should be avoidable in a well-managed unit. Starving and crushing are related, since an injured piglet cannot compete effectively for milk. Equally, a poorly fed pig is physically weak, and more likely to be lain on by a clumsy sow.

To reduce the risk of piglets being crushed, the sow is confined in a *farrowing crate* which restricts her movements but is wide enough to allow her to recline in such a way as to expose all the teats for feeding. (Figures 11.1 and 11.2 show two styles of farrowing quarters.) The floor should be of a material that is both comfortable and easily cleaned, and capable of being disinfected between each farrowing to prevent the build-up of disease. There should also be an area to which the piglets can have access (*a creep*) where the temperature is higher than in the rest of the farrowing house. This is necessary because newborn piglets have low energy supplies and lose heat very quickly. A temperature of 28–30 °C is needed, which is reduced gradually to 20 °C by the time they are five weeks old.

The first stage in the successful rearing of any animal is to ensure that it obtains colostrum as soon as possible after birth, so the stockman must be present to ensure that all the piglets suckle the sow immediately after birth. (Colostrum has a different composition to the sow's normal milk, being richer in protein and containing antibodies against the herd's infections.)

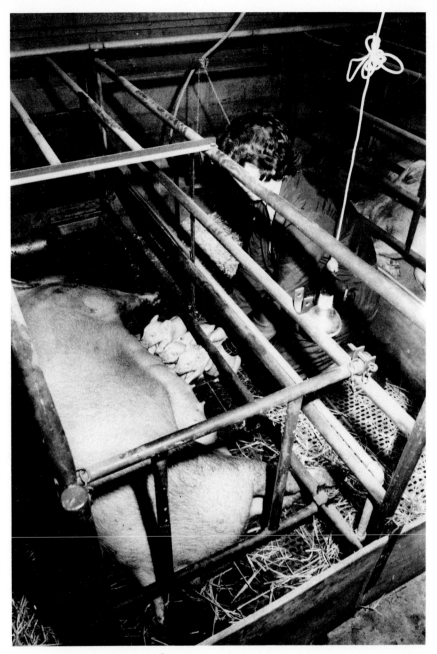

Fig. 11.1 An indoor farrowing crate which prevents the sow from rolling over and crushing the piglets. Here pigs from another sow are being cross-fostered (Farmers Weekly)

Fig. 11.2 An outdoor farrowing hut with a straw-bedded creep heated by gas (Farmers Weekly)

Thereafter, successful natural rearing depends on each pig obtaining approximately 20 ml/hour of milk for at least the first three weeks of life, when it begins to eat solid food. This may not be possible if the litter is large and the sow is producing insufficient milk, or there are more piglets than functional teats. In either case the condition must be recognized early so that the affected pigs can be fostered onto another sow or artificially reared. In order to make fostering easier, some farmers *batch farrow* their sows, and since smaller piglets invariably compete badly for milk, farmers also cross-foster quite large numbers of piglets so that all those suckling one sow are of equal size. (Even then, sow's milk is deficient in iron, so to ensure adequate growth each pig receives an iron injection soon after birth.) The sow's milk yield reaches a peak three weeks after farrowing, after which it must be supplemented with a creep feed which should ideally be increased progressively to minimize the dietary change at weaning. Pre-weaning creep feeds are expensive, but given the very high rate of feed conversion efficiency at this stage of life the pigs should be fed as freely as possible.

In all this the experience and interest of the stockman are essential, and if losses are to be minimized he must have adequate time to observe the pigs and anticipate problems before they arise. To make this task easier, and the stockman's working hours more reasonable, farmers are now synchronizing farrowings and using labour-inducing hormones to ensure that sows farrow during working hours on working days.

If a high survival rate is important to the profitability of the enterprise, the frequency with which sows farrow is equally important. The average gestation period is 114 days. If the pigs are weaned at 1–2 weeks and the sow comes into oestrus five days after weaning, the reproduction cycle then measures 158 days, and it is possible to produce 2.3 litters annually. The actual average for recorded herds in 1979 was 2.2, a significant increase over the figure of 1.8 in 1970. The farmer cannot influence the gestation period, but the period between farrowing and subsequent mating is to some extent within his control. He may also be able to breed gilts at an earlier age, thus ensuring that they produce as many litters as possible in their lifetime. Unlike the cow, the sow will not come into oestrus whilst lactating, so the length of time that elapses before weaning has a determining influence on the onset of oestrus, which occurs some five days after weaning. As long as suitable conditions are provided, pigs can be weaned at any time between ten days and eight weeks old. Eight weeks was the traditional period, but this means that even if the sow is successfully mated at the first attempt, only two litters a year could be obtained. Very early weaning – at two weeks – is uncommon, but 4–week weaning seems to pose few problems on well-managed farms, and allows about 2.3 litters to be obtained annually. This improves the financial return from the enterprise, since it costs the same to keep a sow however many litters and however many pigs she rears. The higher the number reared, the lower the fixed costs of sow-keeping become per weaner. For the same reason gilts should be mated at the earliest opportunity, so that the unproductive rearing period (and therefore the rearing cost) is kept to a minimum. Gilts were traditionally mated at about 220 days of age, but they reach puberty at about 170 days, and it has proved possible to mate them then without detrimental effect. The contribution this can make to the enterprise gross margin was surprisingly overlooked by producers in the past, but given that sows only produce 3–4 litters in their lifetime, gilt litters may represent up to 30 per cent of the litters of the entire herd.

Pig Finishing

As in other livestock finishing systems, the object is to produce an acceptable carcase at least cost. Carcase quality is measured in terms of fatness, weight, and (in bacon pigs) length. Fatness is easily measured in the pig by the depth of back-fat, which increases with the weight of the animal and the level of feeding (Fig. 11.3). As liveweight increases, more food is required to produce one kilogram of weight gain (i.e. feed conversion efficiency decreases), and since 43 per cent of the total cost of finishing pigs is feed cost, this process should obviously be concluded as early as possible. (The cost of the weaner represents another 45 per cent of the total finishing cost, the remainder being largely fixed costs like labour and housing).

What the finisher has to determine is the liveweight at which his margin over the cost of feed plus the cost of the weaner is at a maximum, since this will be the optimum slaughter weight. As in the case of all other livestock systems it is possible to express this in terms of the optimum for the individual animal, or per unit of fixed resource. (For sheep and cattle the measure was *per hectare*, which laid stress on the stocking rate of land. In the case of pigs the appropriate measure is margin *per pig place* in the housing, which lays stress on the rapidity of the finishing process (rate of gain) and therefore the throughput of the house.) The optimum slaughter weight using these two measurements is slightly different, but falls within the bacon

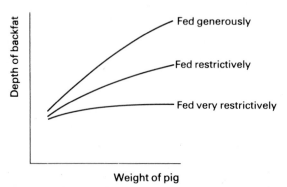

Fig. 11.3 As pigs grow, backfat depth increases (Longman Group Ltd from *Pig production* C. Whittemore, 1980)

pig weight range – around 100 kg, which is the basis for the argument that bacon production is the most efficient pig finishing system. However, it is more difficult to produce bacon pigs to the quality specifications of the curers. It is also becoming more difficult for potential bacon producers to obtain contracts to supply the curers, since the market for bacon is limited and subject to intense competition, which is reducing the number of curing firms in business. Many producers therefore choose to produce lighter-weight pigs for pork, and since it seems that the maximum rate of gain is achieved by pigs growing in the weight range 50–120 kg (when maximum lean growth rate = 0.5 kg/day), there is a considerable amount of leeway. The rate of gain is moreover related to many other factors including the sex of the animal, genetic character, housing environment and the nutrient quality of the feed, and one of the essentials of profitability is to exploit the genetic potential by feeding sufficient high-quality food throughout this period.

The finisher tries to achieve a compromise which results in a system which is economical of labour, ensures efficient utilization of his buildings, and achieves the maximum possible level of feed conversion efficiency *without* loss of quality in the final product. These interactions are most easily illustrated with reference to the feeding regime, which involves three decisions: what to feed, how much to feed, and how to feed.

Unlike cattle and sheep the pig is monogastric, so it cannot be successfully fed on bulk feeds. In commercial units energy and protein are provided in the form of cereals and other feeds (mainly vegetable, but with some from animal sources, e.g. fish meal or milk whey). The pig also differs from the ruminant in that it is not enough simply to ensure sufficient protein in total; the pig must have specific amino-acids (building blocks of protein), especially lysine and methionine. As its weight increases its protein requirement declines, and when it reaches 45 kg the cost of the ration can be reduced by cutting the protein content. The range of possible rations which can provide the necessary level of nutrition is very wide, and the finisher can purchase or mix suitable feedingstuffs to obtain the desired level of output and quality of carcase. The high proportion of total cost represented by feed makes its economical use essential. Pigs grow faster if allowed to eat as much as they like (*ad lib* feeding), and this increases the throughput of the pighouse. As the pig matures, however, its feed conversion efficiency falls and feed is turned into fat rather than meat, and since the carcase price in this case is closely related to leanness, the extra throughput may be achieved at the expense of a loss of quality. Some restriction in the level of feeding is therefore needed, which is generally related to the age of the pigs, though this reduces the rate of gain and also increases the labour requirement. A common solution is to feed *ad lib* to 45 kg, and then restrict the feeding, but feeding practice depends very much on the relative prices of feed and pigmeat, and the latter depends on the intended market (pork, bacon or processing). In some situations *ad lib* feeding will be possible right up to slaughter, whereas in other situations this would be totally uneconomic. The feeding regime is also modified by the genetic character of the pigs, since some are bred specifically for *ad lib* and others for restricted feeding systems.

The feeding system is a major factor influencing the design of the housing and labour requirements. Pig feed can be offered in the form of a dry meal, as cubes, or meal suspended in water. Experiments have shown that wet feeding gives better results both in terms of growth rate and feed conversion efficiency. One experiment showed that the amount of food necessary to produce one kilogram of gain was 3.6 kg of dry meal or 3.3 kg of wet meal. Liquid feeding can also reduce the labour requirement since a mixture of four parts water to one part of meal will flow through a pipeline, whereas the restricted feeding of dry meal or pellets is either labour-intensive or requires expensive mechanical feeding systems. However, liquid feed has to be fed in troughs, which take up room and reduce the car-

Fig. 11.4 A 10-row, 120-metre long finishing house with wet feeding, natural lighting and ventilation (Farmers Weekly)

rying capacity of the house, whereas pellets can simply be spread on the floor. The saving in labour, the enhanced feed efficiency and greater rate of gain associated with liquid feed therefore have to be considered in relation to the reduced number of pig places available.

The construction of finishing houses is often elaborate, and always carefully designed. There are several standard designs on the market which farmers can buy as prefabricated units for erection on their own site (Fig. 11.4). The important design features are that the buildings be warm, well-insulated, well-ventilated and easy to clean and disinfect between batches of pigs. A separate dunging area must be incorporated when pigs are floor-fed, or food is contaminated by manure and wasted (Fig. 11.5). Such units are inevitably expensive, but they can be completely mechanized, thus reducing labour and labour costs. Frequently the house is kept in near-darkness except at feeding times (in order to reduce activity), when the lights are automatically switched on and feed delivered automatically. The stockman's real job in such a unit is to observe the pigs to ensure that they are all healthy, and to ensure that the mechanical systems are all functioning correctly. In the best-run houses the former job is given priority, and high standards of management are achieved, but it is obvious that in less well-run units the element of stockmanship can be much reduced, and the operation become very much a production line process.

Apart from their expense, the main drawback of these finishing houses is

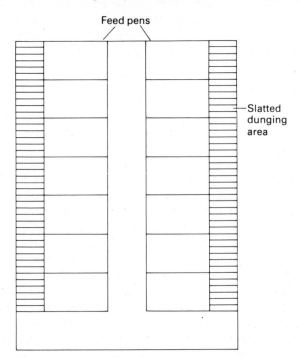

Fig. 11.5 Pens for floor-fed pigs must have a separate dunging area

slurry and air pollution. Slurry is generally stored in a pit under the slatted dunging area which has a capacity of several weeks. It is then removed either by vacuum tanker and transported somewhere for disposal or sprayed onto the land as fertilizer. On a fairly large farm the disposal of slurry presents few problems. The only difficulty arises when the tank reaches capacity in a period of prolonged bad weather, and has to be emptied and sprayed on the land even though the spraying machinery may damage the soil structure. If the pig unit is not part of a large farm, however, the disposal of slurry may be an acute problem. Ideally the pig keeper arranges with neighbouring farmers to spray the slurry on their land, but although pig slurry can be a valuable fertilizer, many farmers object to such arrangements, which can be very difficult to organize on a regular basis. Failure to make reliable arrangements for slurry handling can therefore be a major constraint on the establishment of large pig units on small areas of land. Air pollution can be another serious problem. It may be caused by the housing ventilation system which spreads the smell of pigs, particularly in the summer, or by the spreading of slurry on the land, which has a peculiarly unpleasant smell. Under advanced factory conditions exhausted air from ventilation systems is normally *scrubbed* to remove smells, but this is not generally possible on farms, and research has in fact shown that it is not easy to clean the air from pig houses. Pig farmers seem to acquire an immunity to the smell, but most people object strongly to it, and therefore oppose the location of pig farms in the vicinity of residential areas. The same problem arises with intensive poultry units, and the location of both (and their design) have recently become more carefully regulated by local authorities in order to reduce their potential nuisance to the community.

POULTRY PRODUCTION

If pig production remains an agricultural enterprise at the weaner stage, and finishing has often remained an on-farm activity (albeit industrialized), the same is not true of egg and poultry-meat production. The hatching of eggs and rearing of poults is carried on by a few agribusiness companies, and the majority of egg and meat production is undertaken by another handful of companies. Only a very small proportion of eggs, chicken and turkey is now produced by independent farmers, and even they generally have contracts with one of the large companies. Certainly the traditional on-farm flock of a few hundred hens has virtually disappeared, to be replaced by the unit of a few hundred thousand. (For the sake of concision, the following comments are restricted to chickens, though it should be remembered that turkey production is expanding rapidly, and resembles the production of chicken meat in all essentials. Having been formerly a traditional seasonal enterprise on some farms, where turkeys were reared for the Christmas market, turkey production is now concentrated in intensive units which supply the growing turkey-meat market in the UK.)

As in pig production, the division of labour is well established in the poultry industry. There are large hatchery companies which 'design' chickens genetically for egg or broiler production. They then hatch eggs and sell day-old chicks. Egg producers may rear these day-old chicks to point of lay themselves, or buy in pullets ready to lay, but for broiler production purchased chicks are normally reared and fattened in the unit. The egg producer may prefer to rear his own birds for the same reason that broiler producers rear them: because how they are reared affects future performance. The young birds are also given courses of vaccinations and receive treatments like beak trimming which producers generally prefer to have under their own control. If the birds are to be housed in cages throughout their production life, it is also advantageous to accustom them to the cages from the earliest possible age. (The use of tiered cages obviously makes better use of house space.) Another factor of importance during the rearing process is the lighting pattern to which the young birds are exposed. This affects the age of maturity, food consumption, the body weight of broilers and the number and size of eggs produced by layers. In controlled environment housing it is easy to alter the day length, and the usual pattern is a diminishing day length during the growing stage (down to eight hours at twenty weeks of age) and increasing light in the production stage (up to sixteen hours after thirty weeks). There are many patterns recommended for the different *brands* of chicken, however. (The word *brand* is literally used by the hatchery companies which produce them.) Broilers are also subjected to special lighting patterns which encourage regular feeding and therefore accelerate the growth rate and throughput of the house.

Egg production is concentrated in very large units, with over 100,000 layers typically on one site. The birds are housed in buildings whose total environment is under the poultryman's control. They are well-insulated, have forced ventilation, and day length is carefully controlled. Layers are generally housed in tiers of cages, generally three levels high (Figs. 11.6 and 11.7). The floors of the cages have a slight slope so that the eggs roll into a trough at the front. This contains a conveyor belt which collects the eggs mechanically. Feed is also provided by a travelling hopper which deposits

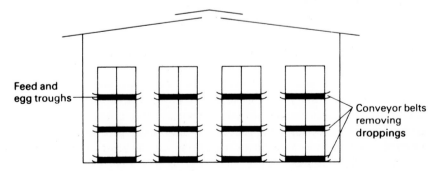

Fig. 11.6 A typical battery house with vertical tiers of cages

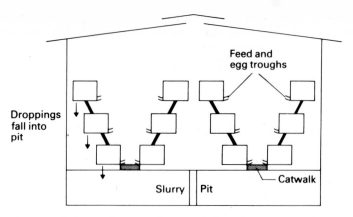

Fig. 11.7 A battery house with stepped tiers of cages over a slurry pit.

food in a second trough at predetermined intervals. Manure collection and disposal cause as many problems as they do in pig houses: a producer with 100,000 layers has to dispose of 5,000 tonnes of manure annually. Collection inside the house is fairly simple. In the original vertical battery house (Fig. 11.6) there is a conveyor belt below the wiremesh floor of each cage tier onto which the droppings fall. Once or twice a day the conveyor is run and the manure is brought to the end of the house where it is scraped off the belt into a storage pit. If the tiers of the battery are stepped, a pit can be made beneath them into which manure falls directly (Fig. 11.7). The manure remains there for the entire length of the flock's productive life (or even several flocks) before being cleaned out. During this period it dries without any loss of its nutrient content, which may be as high as 4.9 per cent nitrogen, 2.0 per cent phosphate, 2.0 per cent potash and 5.0 per cent calcium. Poultry manure is thus a potentially valuable fertilizer, but still difficult to dispose of.

The object of poultry production is again to turn feedingstuffs into eggs or meat as efficiently as possible. In the case of meat, the bird is a most efficient converter, needing about 2 kg of feed to produce one of meat, depending on the age of the bird (since the feed conversion efficiency again declines with increasing age). The composition of the rations of broilers and layers need not be discussed here, especially since many poultry keepers seem to have relatively little knowledge of the nutritional requirements of poultry, and simply feed specially formulated purchased feed at the manufacturer's recommended rates. Provided they do this, however, and follow strictly controlled programmes of health control and management, the operation can be literally reduced to an industrial process which has little in common with agricultural production. If maximum production is to be achieved, there is very little room in this system for error, and very little room either for sentiment, since more than in any other food production process which originated on the farm, poultry keeping has become a matter of economics rather than of husbandry. It is this aspect of intensive live-stock production which has attracted public attention, and aroused debate about the whole issue of animal welfare throughout the agricultural industry to which this chapter's special topic is related.

SPECIAL TOPIC: ANIMAL WELFARE

The existence of caged battery hens has come to epitomize the problem of animal welfare on the farm, even though, as this chapter explained, such conditions are rarely to be found on farms. Animal welfare is a very conten-tious subject. It is also one which is very difficult to study because like hu-man well-being it cannot be measured in any objective, scientific manner. The question of animal welfare arises not just in relation to intensive live-stock production, but in all farm situations, during transportation and hand-ling from farm to slaughterhouse, and in the abattoir up to the point of slaughter.

All of these have been studied over the years, and in most countries reg-ulation or codes of practice exist which govern more or less extensively the practices that can be employed. Paradoxically, the farm situation was the last to receive the attention of the legislator. Abattoir practices were sub-jected to legislative control in the nineteenth century in Britain, and though the legislator has been less directly concerned with the handling of animals

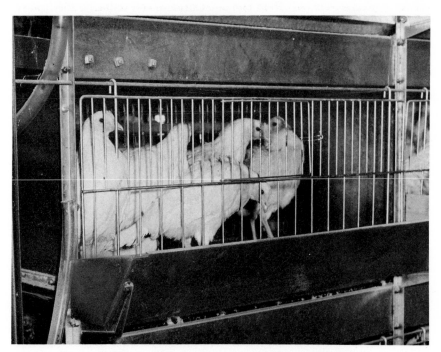

Fig. 11.8 Caged pullets in a battery house (Farmers Weekly)

during transportation, various quasi-official as well as voluntary organizations have concerned themselves with their welfare. There is no doubt that removing animals from a known environment causes stress, and while this is inevitable, there is clearly no justification for causing more distress than is absolutely necessary. This means taking care to reduce rapid changes in temperature, and care is necessary during loading and unloading to prevent animals from slipping and damaging themselves. (In one recent study 37 per cent of the animals arriving at an abattoir had received bruising.) Considerable attention has been paid to this issue as a result of the long distances over which livestock is now commonly shipped in the EEC. Journey of 1,000 or even 1,500 km are not uncommon, and if distress is to be prevented provision obviously has to be made for rest, watering and feeding during the journey. An EEC directive exists which lays down acceptable standards for animals traded within the Community, and many other countries have similar regulations, but there is evidence that they are quite commonly disregarded. In fairness to most farmers it must be said that the real offenders are transport contractors, but a responsible farmer should not leave his animals in the hands of any contractor whom he has reason to suspect will not treat them well.

The welfare of animals actually on the farm only came in for scrutiny following the growth of intensive livestock systems. These are designed to achieve high levels of feed conversion efficiency, as we have seen, and this means protecting animals from climatic extremes and reducing their movement so that energy is not 'wasted'. The high cost of providing a controlled environment entails high stocking densities, and this reinforces the case for confinement. Confinement is not necessarily detrimental to animals. Confinement in a building provides shelter and protection against predators; it may reduce the incidence of bullying among animals, and as in the case of the farrowing crate, confinement may prevent one animal from damaging another – even its own offspring. The confinement of bulls is also generally held to be necessary to protect farm workers and the public at large. The problem is thus not confinement as such, but the degree to which the animal's behavioural urges are frustrated, which depends on the closeness of confinement and its form. In practice this means giving attention to the size of battery cages and the number of birds they may house; the size of crates used for rearing veal calves; the length of time which sows spend in farrowing crates, and how they are kept between farrowings. The structure must also be considered, with special emphasis on ventilation, and even the kind of illumination installed. The floor material and bedding are very important for the animal's comfort, and of course the care with which housing arrangements are cleaned affects very materially the health of the stock.

There *are* methods of confining animals which cause suffering, the most obvious of which is the use of crates for veal calves which are so small that the animals cannot even turn round. The ultimate answer to these rare instances of animal ill-treatment must be legislation and effective surveillance.

More difficult to deal with is the problem of ill health, which is almost certainly the major cause of animal suffering on the farm. In this respect, paradoxically, intensive units may well be better for animals than open farm situations, for strict programmes are routinely implemented to safeguard animals' health, which is in any case protected against the normal hazards of the natural environment. Physical ill health caused by parasites, especially in pigs, is significantly reduced in intensive units (though if the ventilation is inadequate, pneumonia or other bronchial conditions may result). Mental health, if the term may be used, may nevertheless be worse in such conditions. It is certainly established that intensive conditions reinforce animal *vices* – tail biting in pigs, feather pulling and even cannibalism among chickens, but this may simply be the result of boredom. Such vices result in obvious damage to the victims, but some of the 'cures' offered are equally undesirable. Reduced lighting is perhaps an acceptable means of reducing animals' attacks on each other, but mutilation is also used. Tails may be docked in pigs, and the upper mandible of birds is often reduced by approximately a third (misleadingly called 'de-beaking'). The recommendation has been made that both practices should be proscribed, but no action has yet enforced the recommendation.

In open farm conditions, poor physical health may well be the major cause of distress to animals, and results generally where farmers disregard the wisdom of calling regularly on the services of a vet, but rely instead on their own diagnostic powers and often ancient remedies. Poor nutrition is another factor which affects animal welfare, and most farmers know this and try to provide as high a level of nutrition as possible. Fortunately good economics as well as humane considerations reinforce the wisdom of feeding animals well. In intensive units some nutritional problems have been encountered: copper poisoning in lambs being reared intensively, and some liver disorders in barley-fed beef, for example. These were the result of insufficient knowledge, however, and once the solution was found the problems ceased. In the production of quality 'white' veal, the calves' nutrition is deliberately controlled to prevent their flesh from developing its normal red colour. The diet is a liquid milk substitute containing as little iron as possible, and the calves have no access to fodder from which iron is normally obtained. There is considerable disagreement as to whether suffering is actually caused by such a diet, but when the feeding regime is considered together with the system of confinement in very small crates, many observers conclude that the entire system is inhumane.

All management of animals, it has to be acknowledged, involves interference with their normal behaviour patterns, and in that sense all livestock farming reduces their well-being. This said, most people would argue that as long as animals are kept for man's benefit, they should at least be kept in the best possible conditions and not subjected to unnecessary suffering. For most people this tends to mean that whereas normal farm conditions are acceptable, intensive livestock rearing should be discouraged as far as possible. However, there is an economic price to be paid for a move away from intensive pig and poultry keeping. In 1980 evidence was presented by the industry to a Parliamentary Select Committee on Agriculture which suggested that egg prices would rise by 80 per cent if poultry keepers were obliged to adopt free-range production methods. Prices would rise by 16 per cent if the number of birds per cage had to be reduced from five to three. Giving evidence to the same committee in 1981, representatives of consumer organizations said frankly that most consumers are more worried by high food prices than by animal welfare, and that few would be prepared to pay higher prices for happy hens.

In the circumstances, the best way to ensure that animal welfare is not needlessly neglected is to encourage the development of high levels of management skill. On the ordinary farm, it is not an exaggeration to say that few animals are deliberately mistreated, and that most livestock men are attentive to their stock's wellbeing for economic if not for compassionate reasons. The main thrust of public concern should therefore be to ensure that the industrial units employ only properly trained stockmen, and to encourage rather than bludgeon farmers everywhere to maintain a high standard of animal husbandry.

12

Agriculture and the market

Agriculture is increasingly a supplier of raw materials for a complex production process which transforms farmgate products into food and other products purchased by consumers. When it leaves the farm, agricultural produce enters the marketing channel which *adds value* in the eyes of the consumer by transforming what would in many cases be unsaleable products into desirable purchases – livestock into packaged joints of meat, wheat into bread and cakes. So much value is indeed added that the farmgate price represents on average less than 50 per cent of the retail price in the UK, and for some highly processed products it is less than 10 per cent. The point is that consumers do not just buy nutrition when they buy food; they buy nutrition plus services, and it is these services which are provided by the marketing channel. Some of the services are real in the sense that the product has been stored, cleaned, preserved, transported, etc., but others add less tangible attributes like prestige and fashionableness which are nonetheless valued by consumers. The combined function of agriculture plus the marketing channel is to provide the correct mix of goods and services to meet consumers' requirements. In doing so the firms and individuals involved must make sufficient profits to provide income on which to live, and capital for investment in new facilities and research into new products which will form the basis of future profitability. It is thus in the interest of the consumer that the firms providing goods and services should be profitable, and it is in the interest of the firms to ensure that the right goods and services are provided at the right prices.

If this mutually advantageous system is to operate effectively, a number of functions have to be coordinated and institutions collaborate, and it is this interacting set of institutions collaborating to perform a wide range of functions which is called the marketing channel (Fig. 12.1). It is literally a communication channel which receives inputs from farmers (and ancillary suppliers of packaging materials, etc.) and sells output to households, and also collects information about consumer needs, or the needs of other components in the system, and transmits them to the agents who have to act upon them. The channel also moves money and risk between the multiple participants. The efficient performance of these various roles is not only important to the participants in the channel; it is also keenly watched by gov-

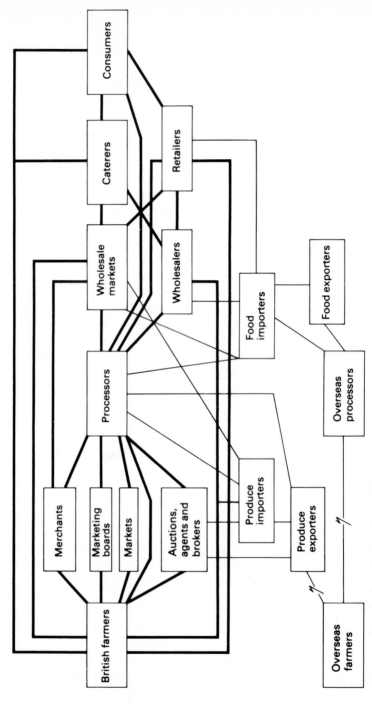

Fig. 12.1 The food marketing channel in the UK

ernments in all countries which intervene if its performance falls below their expectations, either operationally or in the relative fairness with which different participants are treated. If one party is felt to be discriminated against, governments intervene to regulate markets, and farmers, we have seen, have consistently received protection on the grounds that they have insufficient power vis-à-vis the rest of the marketing channel. More recently consumers have enjoyed increasing protection for the same reason, but procedures exist (like anti-monopoly legislation and investigation procedures) which seek to ensure that no single component in the system can unreasonably manipulate it to its sole advantage.

The marketing channel is thus a set of interrelated functions whose detailed structure and operation change constantly, though there is a recognizable overall structure which does not change materially over time. The changes arise from a mixture of causes including changing technology and changing consumer wants which entail changes in production right along the channel and in the agricultural system which provides the raw material. Consumer wants change for many reasons including fashion, but rising incomes in the west have been the main stimulus to changes in the food marketing channel in the last half century. Rising incomes tend to change styles of living which change eating habits, and the food industry not only responds to such changes – it anticipates and accelerates them, developing new products to appeal to new life styles. Research into consumer needs is therefore at the heart of marketing activity, and only when these needs are identified should production ideally begin.

Much of this research is conducted by companies in the food industry on their own behalf since they are large enough to afford the high costs of research and development, and will of course directly benefit from the results. The problem is that relatively little of this information is transferred back to agricultural producers in a meaningful form. Since the majority of food purchased has received some processing, consumer requirements are meaningful to the farmer only when they are expressed in terms of a clear production specification which relates to the raw product he provides. In many cases the consumer's requirements are precisely identifiable because the consumer is a manufacturer who states down to the last detail what he requires of (say) a sample of wheat for baking or a sample of barley for brewing. For certain undifferentiated products like meat, however, the requirements of the final consumer or the intermediate processor are only imprecisely communicated to the farmer. How fat is too fat, for instance? Until carcase fatness is translated into a feed specification and slaughter weight which farmers can be persuaded to adhere to, it is relatively meaningless to urge them not to produce overfat animals which the market does not want. Research in this area has, however, been less intense because the firms involved in the meat industry have been too small to afford the exercise or derive enough individual benefit from it to make the expenditure worthwhile. In such a situation government intervention is again quite common, and can be very useful. In association with representatives of all sections of the British meat industry the government established a semi-autonomous body, the Meat and Livestock Commission, to conduct research and carry out development work in the attempt to translate research findings into action. The Eggs Authority and Home-grown Cereals Authority are similar bodies which have wide research and market information duties as well as responsibility for other related market activities. None of them replace the private companies which obviously continue to do research on their own behalf to further their own share of the meat, eggs or cereals market, but such agencies do ensure that there is some coordination of research, and interpreting it in terms meaningful to the agricultural producer is one of their main functions.

Once consumer requirements are known, they have to be transformed into products and services, and it is the sheer variety of these which makes the agrifood industry more complex than most others. Many problems arise also from the sheer size of the industry, the number of separate production units and the number of products involved, but the nature of agricultural produce increases the complexity of the system. Much of it, for instance, is highly perishable in its natural state. It therefore has to be moved rapidly from farms the moment it is ready, and shipped either direct to housewives for consumption in its prime state or to processors for transformation while it is still in peak condition. A rapid and efficient transportation system is therefore required, which may be easy and fairly inexpensive to arrange if production is concentrated near consuming centres. If production is more remote, however, really rapid transport may have to be provided, like air freight for fresh flowers. Where production is dispersed over a whole country, and is also consumed by every household in the country, the problems of collection and distribution become very great and consequently expensive. Milk is an obvious example, which in most developed countries is handled by cooperative organizations or other agencies which collaborate to operate zoning schemes which reduce transport costs to a minimum. Most milk is still collected individually from farms, and the same is true of grain. Some livestock is also collected from farms, but the collection of one or two animals from each farm is basically uneconomic, and individual animals have long been grouped for onward transport to slaughter or processing centres. These centres developed in association with the railway system in the nineteenth century, but even today the auction marts which serve them act as physical assembly points and price-determining centres where animals change hands.

Transportation is a vital function throughout the marketing channel, and changes in transport facilities and costs affect the location of production, processing and distribution functions. Before the advent of rapid transport systems milk production was located mainly in the immediate vicinity of population centres; before the advent of efficient refrigeration, slaughterhouses were located near consuming centres and animals were shipped there for slaughter. Today abattoirs can be located in the production areas, and carcases or even individual packs of meat shipped direct to retail outlets. The structure of retail trading has itself been dramatically influenced by the increasing mobility and willingness of consumers to travel substantial distances to do their food shopping. The distances involved mean, however, that food shopping is done less frequently than it was, and more goods are purchased at one visit. This has meant in turn that the stocks of goods held by shops have increased, and more capital is tied up in those stocks. The range of goods stocked by large supermarkets and hypermarkets has also expanded, answering the public's preference for one-stop shopping.

The holding of stocks is a major function of the marketing channel at all its levels, not just at retail level. The housewife is an irregular purchaser of food, who herself holds stocks, but like the retailer she generally holds them for only a few weeks at the most, whereas the channel holds long-term stocks. Many agricultural products have a marked seasonal pattern, we have seen, and they therefore need storing and releasing onto the market gradually to meet consumer demand. The storage agency differs from country to country and commodity to commodity. In the USA, for instance, grain is stored by private elevator owners; in Britain it has traditionally been held by farmers, and on the Continent the service has often been provided by agricultural cooperatives. If the producer stores it, he needs a source of finance to provide income and pay for the inputs for the next crop, and if others store it for him, they need funds to pay him. In addition, the stored produce has to be insured against physical loss, and often it is also insured against the risk of a price fall while it is in store. In most countries institutions exist outside the marketing channel which provide these *facilitating services*. Where agricultural produce cannot be stored in its original form, but has to be processed for storage (e.g. milk powder or frozen vegetables), the stocks are held by food processing firms, and they too require sources of finance.

The transformation of agricultural produce into consumer products is a major function of the marketing channel. The form of the produce may be changed for a variety of reasons, one of which may be simply to preserve it during storage. Many products are not usable in their raw state, however. Oil from oilseeds has to be extracted and converted into margarine or cooking oil; livestock has to be slaughtered and transformed into cuts of meat or fabricated products like sausages and hamburgers. In these cases the final consumer is concerned only with the quality of the final processed product, and it is the processor who translates this into a quality specification for the raw ingredients, but where the consumer buys fresh farm produce which has simply been packed, she is directly interested in the quality of the produce. In all cases, it has been well established, what most consumers value most highly – be they housewives or food processors – is consistency. It may be consistently *high* or consistently *low* quality (for the level of quality required is affected by the use to which the produce is to be put, the income level of the household, etc.), but what most consumers want is that the quality should be consistent over time.

This can be assured by processing and branding which standardize a given product (so that Brand X cornflakes always taste the same). This is the food industry's principal means of stabilizing its market, and provided the final product is recognizably the same it can adjust its processing techniques within limits to counteract differences in the quality of its raw materials. In the case of bread, for example, it is possible to use different combinations of wheat varieties to achieve a constant protein level, which is essential to guarantee a standard product. This does not mean that bakers are uninterested in protein content and quality, for they in fact select wheat in relation to these very criteria, paying a premium for samples with the correct balance. When they cannot obtain such samples, however, they can compensate for any deficiency and still produce a standard article by blending ingredients of different quality. The quality of such products is thus founded on the measurable quantity of specific ingredients important to the processing activity, and this is true of many raw materials which undergo processing. The fat content of milk is precisely measurable, for instance, and affects the processing use to which it can be put. The percentage of sugar in sugar beet is vital to the refiners, and the producer is paid on the basis of this very precisely determinable measure of quality. For most produce which goes for further processing there is invariably a minimum quality level below which it is unlikely to be purchased by anyone, and in the past it was true that a proportion of farmers were satisfied if they simply produced a product which met this minimum standard. The growing power of the food industry and increasing discrimination of the consuming public are nevertheless making all farmers more quality conscious, and the closer producer prices are related to quality requirements which farmers can realistically meet, the more will this be so.

This is particularly important where deficiencies in produce cannot be compensated for in the processing. There is no way in which hard peas can be satisfactorily tenderized before freezing, for example. Quality deficiencies also cannot be corrected, of course, if produce is sold to the housewife in much the same condition that it left the farm. In this case, all that can be done is to separate the different qualities and offer them separately, which is the object of classification or grading. The difference between classification and grading is that the former simply indicates two distinct qualities without saying that one is better or worse than the other. Two cauliflowers of equal freshness but unequal size suit different-sized households, and to classify them by size is useful to the greengrocer, who can order a mix which he knows will suit his regular customers.

The criteria used to classify or grade agricultural produce are a mixture of objective quantity measures and subjective quality measures. Eggs, for instance, are classified according to their weight, which is purely quantitative, but they are also sorted by their colour (brown or white), which reflects a consumer preference which is not strictly objective but nonetheless real. Cucumbers are classified by length, straightness and colour, development and absence of damage. These consumer standards are based on characteristics which are known, or which are believed to be important to consumers, though the evidence is that often they reflect what the organization which operates these standards *thinks* consumers *ought* to want. A good example is the size of apples, which the trade insists must be above a certain size because consumers like large apples. The evidence is that mothers very often express a preference for small apples, since children waste less of them than they do of large ones, but this detail is ignored by classification schemes which dictate that the largest apples command the highest prices.

It is obvious to the consumer that classification or grading are not standard practices for all agricultural produce in the UK. In the USA, grades (not just classes) exist for virtually all agricultural produce, and are marked clearly right through to point of retail sale. In Britain, even those schemes which exist are not generally carried through to the consumer level, stopping instead at the wholesale/retail boundary. For example, a classification scheme exists for meat, a product for which it is particularly difficult to achieve any consistency, but it was designed to help butchers obtain consistent carcases, and to indicate to producers through a system of premiums and penalties the kind of carcase they should be producing. The classification scheme is based essentially on two characteristics: the amount of fat on the carcase (since it is well established that the average consumer does not want fat meat) and the carcase conformation, which gives an indication of the relative proportions of various meat cuts. Fairly complex scales of measurement have been devised by the Meat and Livestock Commission which allow carcases to be very closely classified by an inspector in the abattoir, but the scheme is not compulsory, and very little meat is actually classified in this country. In the USA, by contrast, meat grades are used for all meat purchased by consumers which are marked right through to the supermarket, and it is difficult to sell consumers meat that does not carry these well-known grades.

Experience in the USA has shown that such a scheme can benefit the marketing channel as a whole – producers, intermediaries and consumers – by ensuring that it does act as a channel for the communication of market needs back to the producer, who can change his system in order to produce a better article for which he receives a higher return. In Britain, however, there is little evidence that this is happening. The meat classes have not become the basis of trade at the wholesale level because traditional trading habits have proved a strong obstacle to the new system. Price differences therefore reflect quality differences very imprecisely, and have had very little effect on producers as a result. This is not just the fault of the meat trade. As Chapter 3 argued, it is also a consequence of the income support policies which have cushioned producers against market forces and guaranteed prices without making the quality of produce a high priority, and only

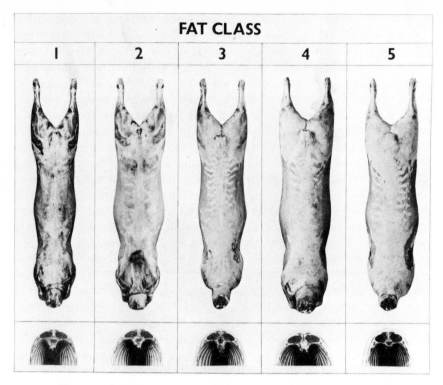

Fig. 12.2 Lamb carcases classified by fatness. Class 1 is excessively lean; Class 5 excessively fat. The market requirement is for Classes 2 and 3. Despite the existence of this scheme, however, only 10 per cent of sheep and 20 per cent of beef carcases are on average classified (M.L.C.)

a change in this respect will make any real impact on producers. It is at least arguable, however, that a classification scheme which did continue through to the consumer level would lead to wider adoption of the existing guide-lines, and allow the marketing channel to fulfil its basic communication function more effectively than it does at present.

The most obvious function of the marketing channel is to exchange goods. At its simplest, the exchange is a direct transfer between the producer and the consumer – say in the farm shop, or at a roadside stall. However, this is impractical for all but a minority of transactions, given the existence of some 250,000 producers, some 18.5 million households, and the infinite number and variety of agricultural products and processed foods which consumers require. In addition, the great distance both geographically and in time between production and consumption means that efficiency is actually increased by the existence of large numbers of intermediaries who transport, store and process products, and are ultimately responsible for determining their price. Each party to a sale transaction obviously tries to influence the terms of sale to his own advantage, and where the housewife deals directly with the producer, both may well feel that they have more say in determining a price which is fair to each. In dealing through intermediaries both feel powerless to determine a 'fair price', and it is this feeling which has encouraged the revival of direct encounters at the roadside stall. For most of its food purchases, however, the majority of the consuming public will continue to be dependent on the intermediaries in the marketing channel, and the consumer's desire to obtain reliable supplies at reasonable prices will be more realistically met by the presence in the chain of *many* intermediaries, each trying to influence the terms of sale to their own advantage.

The manufacturer stresses the advantages of his product over those of his competitors; the retailer stresses the convenience and pleasure of using his shop, or the lowness of his prices. Where there are many manufacturers and many retailers the consumer is able to choose between them, and there is competition for her business, resulting generally in an improved standard of service. From the producer's point of view it is equally desirable that there should be many buyers competing for his produce, but it is a fact that in many sectors of the industry the number of buyers is limited, and farmers have always felt that the terms of trade are therefore manipulated to their disadvantage. The suppliers of farm inputs are also often restricted in number, so farmers may be weak buyers as well as weak sellers, and it has often been thought that they need to develop *countervailing power* to improve their terms of trade. This view has been shared by governments as well as farmers, leading to the creation of institutions which exist only in the marketing channel for agricultural products, whose function is mainly to buy or to sell on behalf of farmers. Many of these organizations are producer-inspired and producer-controlled bodies which assemble produce and thereby attempt to increase their power to influence the terms of trade. The simplest such organization is the agricultural cooperative society, which in most countries enjoys a privileged status vis-à-vis other private companies, and in some has transformed the farmer's situation in the marketing channel.

These collective organizations can offer services to potential buyers which it would be uneconomic for individual farmers to offer (like the sorting and grading of produce, or some primary treatment). They can achieve economies of scale in certain activities (for example, storage and transport), which means that the function is performed at a lower total cost than if it were performed by individual farmers or the purchaser, and this should result in a higher producer price. The offer of such services obviously gives the cooperative some extra bargaining power with potential buyers, but the real advantage which most farmers seek to achieve by belonging to a cooperative is the ability to *determine* the price at which they sell. Like the manufacturers of industrial goods they want to become price *makers* rather than price *takers*, and at its most extreme this expresses itself as a desire to be able to cut off supply, in much the same way that a trade union calls a strike, in order to force up prices. Except on very rare occasions, however, such action has not been taken, the main reason being the ultimate lack of cohesion among farmers, though such action has also been undercut by the ability of buyers in the long term to find substitute products or substitute suppliers. (An attempt by the cotton growers of the USA to force up the price to textile manufacturers, for instance, resulted in a massive research programme to develop synthetic fibres which completely undercut the cotton market.) In less extreme circumstances, however, agricultural cooperatives in many countries have significantly changed the marketing of farm produce, though some would argue that this has not been to the overall advantage of the marketing system.

Agricultural cooperatives in developed countries are voluntary organizations financed and controlled by their farmer shareholders, with the intention of providing services to members at cost. (In developing countries there are 'compulsory' cooperatives to which producers must belong, which are very largely controlled by governments, though an appearance of producer control is often preserved.) The philosophy of the cooperative society is thus fundamentally different from that of the private company, whose first responsibility is to return a satisfactory profit on operations in order to remunerate its shareholders, who may change from day to day and have no direct interest in or knowledge of the company's business. The object of the cooperative is not to maximize profits, but to provide services to members, and though in practice many cooperative societies are run in virtually the same commercial manner as private companies, this fundamental principle must be respected. Many cooperatives are run in a strictly commercial manner because they are in direct competition with private firms for farmers' trade. They supply inputs as well as market members' produce, and their presence in an area may certainly improve the terms of trade available to farmers. If the coop is as efficient as a private firm, and does not have to make a profit on operations to remunerate outside shareholders, it clearly ought to be able to charge lower prices for inputs and pay higher prices for output, since it only has to meet its operational costs. The rational farmer would therefore become a member of the cooperative and its strength would grow, while its private competitors would be obliged to adjust their prices to meet the competition. In practice this does happen, but in the process, *all* the farmers in the area of the cooperative, not just its members, benefit from the increased competition.

This raises a problem which applies equally to all voluntary organizations (including trades unions, chambers of commerce, etc.): namely, that many more people benefit than those who support and fund the organization. In the case of the farming cooperative the solution to this *free rider* problem is relatively easy. The member is charged the same price for purchases as he would pay a commercial firm (or receives the same price he would receive for produce), and the difference is returned in the form of a bonus at the end of the year. In this way, only members benefit from the coop's activities, and there is an incentive for non-members to join.

All cooperatives face many problems both at their establishment stage and during their later development. They must be efficiently managed (so the best-managed employ professional managers who know the marketing channel), and they must be substantial enough to remain competitive with the private firms in their area. In some parts of Europe, marketing coops have become so efficient and substantial that they have arrived at an understanding with their former private competitors to share out the trade in a given area. Some milk marketing cooperatives and private dairy firms have for example divided whole areas into zones in which each has a monopoly of milk collection and marketing, in which case it is questionable whether the 'coops' are any longer voluntary organizations, since a farmer living within the area of a given cooperative has no alternative but to sell milk to it. Where cooperatives are still in competition with private firms, the loyalty of members must be the first concern. Most farmers, it must be admitted, judge the efficiency of any organization – including their cooperative – on the prices it charges and pays, and it frequently happens that a coop does not offer the best price. In this situation farmers would ideally remain loyal to their organization, and very many do, but it is not uncommon to find members of coops trading elsewhere. Many marketing coops therefore have written agreements with their members to ensure that supplies can be relied on to meet orders, without which they could never build up a trusted marketing organization.

The question of loyalty becomes even more crucial when attempts are made to restrict supply in order to raise the price. A cooperative may attempt this either by restricting membership, or by restricting the production of each member (or a combination of both methods). The restriction of membership is not generally believed to be consistent with cooperative principles, however, and the deliberate exclusion of farmers who are otherwise eligible means the loss of the cooperative's privileged status in many countries. Where it tries to restrict members' production a cooperative faces the greatest test of member loyalty, since if the coop is successful in raising prices there is a very great incentive for members to seek ways of circumventing the production control in order to obtain the higher price on as high an output as possible. They may be assisted in this by buyers who wish to undermine the cooperative so that they can obtain supplies at lower prices, and the threat which this represents has generally been such that governments have again intervened to reinforce voluntary agreements. In the USA marketing agreements and statutory marketing orders are used to support voluntary efforts, and in many developed and developing countries of the Commonwealth the solution which has been adopted is the *marketing board*.

An agricultural marketing board is engaged in many aspects of the marketing of a particular commodity, but all the producers of the commodity must this time register with it, and are subject to the rules and regulations it makes. There are large numbers of marketing boards; in Canada alone there are approximately 100. The most important distinction between them is whether they physically trade in the commodity concerned, or simply regulate the trade to some degree. In Britain there are examples of both types. The five regional Milk Marketing Boards and the Wool Marketing Board are normally engaged in physical trading, whereas the Potato Marketing Board only acts as a buyer of last resort in years of surplus, and normally only regulates the trade. Trading boards actually buy and sell the commodity. The MMBs have the right and duty to purchase all milk from dairy farmers and sell it to wholesale dairymen. Some of these are commercial dairies which treat, bottle and sell milk to retail outlets or direct to consumers; others are manufacturers of milk products – butter, cheese, dried milk, chocolate, etc. This processing may be totally independent of the Boards, or it may come under a processing division of the Board, in which case it is responsible for treating all processors equally fairly. The Boards must accept all milk offered to them, and cannot operate any control on the quantity produced. They must collect milk from producers and pay everyone the same price for the same quality of milk, so that no producer is discriminated against. (There are some slight regional differences in price and transport charges, but they are minimal.) They do not charge the same price to all purchasers, however, charging a different price according to the use to which the milk is put and the realization price of the final product. It is clearly in the Boards' interest to sell as much milk as possible in the highest-priced market, which in Britain has always been the liquid market. The revenue is then *pooled*, and the producer is paid this pool price per litre. The Boards also have a statutory responsibility to ensure supplies of milk for the liquid market, and as Chapter 8 noted, to this end they encourage winter production by paying a premium for winter milk.

The MMBs (and the Wool Board) thus have a monopoly in the purchase of their commodity which is authorized by legislative statute. This status was only confirmed, however, after it was shown by a poll that a majority of producers were in favour of it. Thereafter all producers were obliged to accept the majority decision and the statutory nature of the Boards' existence. The Boards' status was challenged by the EEC Commission in the late 1970s, when another poll of milk producers produced a massive majority in favour of their preservation. The EEC's objection to their existence was that the marketing boards are not exclusively producer organizations, as are agricultural cooperatives on the Continent, which is undoubtedly a fair criticism. The British government intended these boards not only to safeguard producers' interests, but also to improve the efficiency of the whole marketing channel and to ensure that consumers' interests were not damaged. The legislation regulating their composition therefore requires that although the majority of the management board shall consist of producers' representatives, there must also be some Ministerial appointees whose duty is traditionally to ensure that the interests of the rest of the marketing channel are not sacrificed to producer interests. In voting for the continuation of the MMBs, milk producers thus voted indirectly for a unique organization which tries to reconcile rather than oppose conflicting interests in the marketing chain. It is thus unlikely that marketing boards could act to control supply in such a way that it damaged consumer interests, whereas the same is not true of the cooperatives which represent producer interests to the exclusion of consumer interests in Continental Europe.

Marketing boards in this country and elsewhere have been active at the farm level, in processing and wholesaling, and at the consumer level of the marketing channel. They have rationalized transport and warehouse arrangements to reduce costs. They have tried to improve farm practices in order to improve the quality of produce (for example, improving the shearing and handling of wool). They have introduced quality-related pricing systems to encourage farmers to produce the right product, and in association with other members of the channel they have been active in promotion efforts to enlarge the total market for a commodity. It is impossible to estimate precisely what their contribution has been to the improved efficiency of the market, since it is impossible to guess what would have happened without them. Their critics have argued that in countries where they do not exist producer cooperatives have achieved similar benefits, but it is arguable that the representation of consumer interests guaranteed in their constitution is a distinct advantage which producer-oriented cooperatives do not have. There is no doubt that the marketing boards' principal function has always been, and still remains the protection of the producer, but it is not true that this has led them to encourage over-production or a waste of resources. As noted in Chapter 8 for example, it is possible to argue that more milk is produced in the UK than is needed, and is subsidized out of general taxes, but this is not the fault of the MMBs, which do not determine the milk price. It is the government which fixes the producer price that they administer, and of course, there is over-production elsewhere in the EEC where there are no marketing boards for milk.

Many producers maintain that neither marketing boards nor cooperatives have succeeded in increasing the farmer's return, by which they mean that they have not increased the farmer's share of the consumer price of food. Farmers' organizations constantly stress that the farmer receives on average less than 50 per cent of the consumer price, and that this figure has declined rather than increased. This has happened everywhere in the developed world, however, and it is less the result of exploitation or marketing inefficiency than of rising costs.

The difference between total consumer expenditure on food and the revenue received by farmers – the *marketing bill* – is a payment to other agents for services rendered. If the number of services increases, or the cost of providing them increases, the size of the marketing bill also increases. Equally, if a service formerly provided by a farmer is now provided by a marketing firm, the farmer's share will decline and the share received by the marketing intermediaries will increase. Conversely, if the housewife provides a service for herself for which she used to pay, the total size of the marketing bill will decrease, because housewife services are not costed. In fact, it appears that all these changes have taken place to a greater or lesser extent. Consumers now serve themselves in shops; farmers no longer make butter, and may not even grade vegetables, employing others to do it for them. Consumers are certainly purchasing more convenience foods, and the service content of most foods is rising. Finally, and most important, the cost of providing these services is rising faster than raw material prices. Many marketing services are by nature very labour-intensive, so labour costs are a major component of all marketing costs, and labour costs have risen rapidly in the last thirty years. The other major costs are land and building, reflected in rent or interest charges, and costs of fuel and packaging material (much of it produced from and using non-renewable energy resources). These costs have also risen in most countries much faster than agricultural output prices, thereby reducing the farmer's share, and it is not true that profits in most sectors of the marketing chain are as high as might be expected if farmers were indeed being exploited.

The cost of marketing could be reduced by reducing services, or by becoming more efficient, thus reducing the cost of providing the same services. As in agriculture, efficiency in the marketing channel is increasing, for processors, retailers and others are as keen to exploit economies of scale and substitute capital for labour as are farmers (and there is evidence that their gains in efficiency have been as large as those made in agriculture). It is also not clear that there are many services that consumers are prepared to do without. Advertizing has been criticized as one marketing activity which uses resources unproductively, and the services of wholesalers and other *middlemen* are frequently dismissed as unnecessary or unnecessarily expensive. It may certainly be true that there are too many links in a marketing chain, and that a rationalization would reduce costs, and it may be in that direction that farmers should turn to increase their share of the marketing bill, for they are unlikely to receive a bigger share purely as producers.

The improvement of agricultural marketing has been a watchword in the farming industry for two decades, and what this means to an increasing body of farmers is that they should assume some of the functions which they have lost to other agents, or which they never previously performed. This often means collaborating in groups to provide cleaning, storage or similar facilities, or more ambitious processing plant. There are farmers' cooperatives which own abattoirs and sell meat under their own label, organizing an entire procurement, processing and marketing operation with the aid of professional management, and they are again encouraged in this by government. Organizations like the Central Council for Agricultural and Horticultural Cooperation exist to advise and assist in the formation of such ventures, and to provide financial aid. It may again seem to the non-agricultural community that this is an unwarranted support for the farming industry, but it represents a fraction of the aid offered to other industries, and the return from it could benefit everyone if it achieved the objective of improving the performance of the marketing channel, above all by putting producers more closely in touch with the needs of the market. Nor need the integration of functions be in one direction. Farmers may well have to overcome their traditional suspicion of vertical integration, and also work more closely with marketing and processing companies to establish joint ventures. Given the chronically high cost of land, capital and labour – agriculture's basic resources, the pooling of resources has much to commend it, and the likelihood is in any case that it will increasingly be forced on farmers and firms. They may each lose some independence in the process, but no-one in the marketing channel ever enjoyed independence, so better to make a virtue of a necessity and create a real agrifood industry where the opportunities for coordination are maximized.

13

Conclusion

The preceding chapters have considered agricultural systems as food production systems, and have indicated how economic, social and technical forces affect them and, through them, the food industry. The reason is quite simply that in Britain and most of western Europe agriculture is primarily concerned with food production, and although a few non-food outputs are produced, these are only by-products of food production systems. The obvious example is wool, but hides and skins also contribute to the income of meat producers, and there are multiple industrial uses for other animal and plant by-products. In many other countries the production of fibres, tobacco and other non-food products is an important part of agricultural output: cotton in the Sudan, Egypt and the USA, jute in Bangladesh, rubber in Malaysia, etc. Nearer home, the production of flax for linen manufacture has been a traditional agricultural crop in Ireland. In Britain and the EEC, however, the largest single non-food crop is timber, whose production represents as great a challenge as that of any agricultural crop, and probably more than most. Two sub-sectors of the horticultural industry are also devoted to the production of non-food items – namely, flowers and pot plants, and the production of hardy nursery stock and other garden plants. The area devoted to these crops is not large, but they are highly labour-intensive, and ironically the demand for these products has shown considerable growth, whereas the demand for raw materials for food production has remained constant. Firms in this sector were also in the forefront of better marketing – establishing garden centres and exploiting the potential for mail order selling and widespread advertising in specialist publications, and if incomes continue to rise still more expansion can be expected. Like horticulture in general, however, this is a very specialized sector which cannot be regarded as part of the mainstream of agriculture, which is why it has had to be neglected in this book, and why its growth will not impinge directly on the future fortunes of agriculture.

The special position of agriculture in the past has been closely related to the fact that it is still, as we began by saying, the only reliable source of human food. Today, however, it is no longer the only industry producing food. In the past decade a substantial industry has grown up which produces meat, milk and other food *analogues* from a variety of non-agricultural inputs. All foods are basically composed of carbohydrates, proteins and fats, and the first synthetic food was a fat substitute – margarine (originally called *butterine*) which could be cheaply produced from oils derived from vegetable sources. More recently research attention has focussed on proteins, with special emphasis on developing substitutes for animal-derived products. The stimulus is twofold. First, food processors aware of consumer fears about the possible health risks associated with animal foods have as always turned their attention to the development of new products before they find their market diminishing. More important, however, livestock production, as we have seen, is a high-cost system largely because its biological productivity is low. Beef production yields about 25 kg of protein/ha/annum; even poultry-meat production only yields 40 kg. By contrast, wheat produces 175 kg/ha/annum, broad beans 350 kg, and soya beans 500 kg. Efforts have therefore been concentrated on using plants as a source of protein from which the food industry can fabricate acceptable food products.

In Asia and Africa it would be possible to market this protein for consumption in its natural vegetable state, but the western palate demands a substitute for meat. The new foods therefore need to resemble meat, and it is this which caused the manufacturers their greatest problems. The chemical processes for extracting protein from soyabeans were developed fairly rapidly, for example, but the difficulty was to find a means of transforming the protein into a fibrous product like meat. The first *textured vegetable protein* did not have the same texture as meat, and was useful only as a *meat extender* used in fabricated products like sausages and pies. It was then discovered that the protein isolate could be extruded as very fine filaments which could be 'knitted' to produce a texture very close to that of meat, and since flavour can be added chemically (flavour is only a chemical in nature) it became possible to manufacture protein steaks tasting like beef, lamb or chicken.

At first sight this substitution seems to represent a clear threat to livestock producers, and that was precisely how they interpreted the discovery. The first product which aroused organized agricultural opposition was not meat analogues, but 'coffee creamer', a substitute for milk which the manufacturers were not eventually allowed to call by any name that suggested it could serve as a substitute for the natural product. However, if these analogues are manufactured from crops grown by farmers, the loss to the agricultural industry is not absolute; rather it is a loss to one sector and simultaneously a gain for another. Soyabean producers in the USA have certainly found it to be a very profitable crop, and the opportunity to grow it (or similar beans) in Europe is being enhanced as geneticists and plant breeders develop strains suitable for growing under European conditions. Such a crop would expand the range of break crops available and produce rotational changes, so arable systems should benefit, albeit at the apparent expense of livestock producers.

Appearances are misleading, however, and it is not the case that livestock producers will necessarily suffer from the expansion of this new sector. The human market for non-animal protein products is small and unlikely to expand fast in the developed countries, but there is a huge market for protein for animal feedingstuffs, so these developments in protein production may reduce the cost of producing meat (and milk) by traditional means. This should then make the real product more competitive in price with its substitutes. Techniques have been developed for the production of protein from single-cell organisms grown on a variety of carbohydrate or hydro-carbon substrates. Bacteria, yeasts and microfungi can be cultured and grown on substrates like natural gas, wood shavings, husks of grains, and other carbohydrate by-products. The organisms grow very quickly and have a high protein content which is fairly easily extracted. The resulting protein is probably perfectly safe for human consumption, but it will be difficult to convince the public of this, so most of it will probably be used in animal feedingstuffs as a substitute for expensive products like fish meal. In this case, the price competitiveness of pig and poultry meat is likely to be improved by the new sources of protein.

A basic axiom of the food industry is that nothing should be thrown away, and it certainly seems that all the vegetable sources of protein might be used literally without waste. Oil would be extracted and used for margarine and cooking fats; the protein would then be extracted and used for the production of vegetable protein for human consumption. Finally the carbohydrate residue would form the substrate on which to produce more protein to be incorporated in animal feedingstuffs. This illustrates one development which is certain to characterize all food production systems, including farm systems, in the future – namely, the reduction of waste and the attempt to find uses for every potential waste product. The meat industry has always attempted to use every part of an animal, and arable farmers have traditionally re-cycled their by-products through livestock to produce meat and manure. At present their principal by-product – straw – is generally burned on the field. This has seemed necessary in the past for disease control, but many farmers are trying to find other uses for it. One is to burn the straw in a furnace and utilize the heat for grain drying, thus reducing the energy bill. Another is to treat it so that it can be used as an animal feed. Such practices are still experimental, but they nonetheless indicate a general effort to reduce costs that is being made throughout the industry.

The most cost-conscious area which is receiving particular attention is the energy requirement of agricultural machinery. Manufacturers and farmers are collaborating to reduce the energy requirement of machinery, and some remarkable advances have already been made. In one trial, using identical standards of measurement, one tractor used 24 per cent less fuel than the worst model in the group tested. On-farm working conditions would modify this improved performance, as would the standard of maintenance of the machinery, but considerable savings are clearly attainable given the new cost-consciousness which economic circumstances are forcing upon the farming industry. The most dramatic effect of increasing energy costs and diminishing energy supplies is likely to be felt not in agriculture, however, but in horticulture. Heating costs represent 27 per cent of the total cost of producing horticultural crops under glass, and a long-term rational plan would be to re-locate the production of crops grown under glass in northern Europe (tomatoes, cucumbers, winter salads, etc.) in southern Europe, where they can be grown in the open. This would reduce substantially the need for expensive glass or plastic growing houses as well as the heating requirement, though some of the saving would be offset by the need to irrigate in southern latitudes. There would almost certainly be an overall cost advantage, however, and such a change may eventually come about. It would nevertheless be vigorously resisted by growers in Britain, Holland, France and Germany, who would lobby their governments tirelessly in defence of their present means of earning their living. Already Holland provides a subsidy on heating fuel for its growers in an attempt to maintain their competitiveness in face of southern European crops, and as we have seen throughout this book, socio-political arguments for support of the local farming population will counterbalance economic pressures for the optimum location of production.

Land will, of course, continue to be lost to agriculture for urban use, and it may become as scarce a resource as energy in some situations. Attempts to increase the productivity of land are therefore bound to continue, but in countries like Britain they are equally bound to be opposed where they threaten to change the appearance of the landscape. The views of the public in this respect are almost certain to compel farmers to give greater thought to the impact of their farming practices on the landscape, but if financial compensation is not available where real loss is suffered as a result of this, the public cannot reasonably expect farmers to sacrifice their livelihood. It is by no means certain that the majority of the consuming public shares the fears of conservationists sufficiently to pay for the preservation of the landscape in higher food prices or through general taxation, and research is needed to answer this fundamental question. Once a clear answer emerged, the farming industry could reorient its priorities as necessary, and it is both inefficient and unfair to leave the question unanswered and simply assume that the vociferous conservation lobby is speaking for the majority of the population. The same argument applies to the issue of animal welfare, for there is evidence enough that provided conditions are not seen as being *too* inhumane, lower prices still carry more weight with most consumers.

Though the agricultural industry will continue to modify itself as it has always done in response to external and internal pressures, it will still resemble the agricultural industry we know. As production continues to increase and relative prices fall, farms will become larger and farmers will show more interest in cost-cutting exercises, but farmers will still continue to try to make two blades of grass grow where only one grew before. In this they will be encouraged by governments and supported by research. The special position they have enjoyed may be partially undermined by the development of alternative food sources, but it will not be eroded completely unless and until these alternative sources can completely supply the world's food requirements.

Further reading

Chapter 1

One of the best sources of up-to-date information about the farming industry is the farming press. The best general source is *Farmers Weekly*, and there are innumerable specialist publications like *Big Farm Management, Livestock Farming, Pig Producer, Dairy Farmer*, etc. Regular perusal of these journals will provide more insights into the agricultural industry than any book. Also useful are the technical bulletins and pamphlets issued by the Ministry of Agriculture, and two major annual reviews of the industry: the *Annual review of agriculture*, HMSO, London, and *The agricultural situation in the Community*, EEC Commission, Brussels.

Since the present book is limited to farming systems in temperate regions, a useful supplement would be **Hans Ruthenberg**, *Farming systems in the tropics* (3rd edn) Oxford University Press, Oxford, 1980. The development of agricultural systems is described in **D. B. Grigg**, *Agricultural systems of the world: an evolutionary approach*, Cambridge University Press, London, 1974. The systems approach can be studied in **C. R. W. Spedding**, *An introduction to agricultural systems*, Applied Science Publishers Ltd, London, 1979, and a forthcoming publication likely to be of direct interest to readers of the present book is **M. Blacksell** and **A. Gilg**, *The countryside: planning and change*, Allen & Unwin, London, 1981.

Chapter 2

Griffiths, J. F., *Applied climatology: an introduction* (2nd edn) Oxford University Press, Oxford, 1976. (Ch. 10 is on agriculture.)
Ministry of Agriculture, Fisheries and Food, *The agricultural climate of England and Wales*, Technical Bulletin 35, HMSO, London, 1976.
Russell, E. W., *Soil conditions and plant growth* (10th edn) Longman, London, 1973.

Chapter 3

North South – a programme for survival, Report of an independent Commission on international development issues (the Brandt Commission), Pan, London, 1980.
Fennel, R., *The Common Agricultural Policy of the European Community*, Granada, London, 1979.
Hallett, G., *The economics of agricultural policy* (2nd edn) Blackwell, Oxford, 1981.
Marsh, J. S. and **Swanney, P. J.**, *Agriculture and the European Community*, George Allen & Unwin, London, 1980.

Chapter 4

Beresford, T., *We plough the fields. British farming today*, Penguin, Harmondsworth, 1975.
Edwards, A. and **Rogers, A.**, *Agricultural resources: an introduction to the farming industry of the UK*, Faber & Faber Ltd, London, 1974.

Chapter 5

Abbott, J. C. and **Makeham, J. P.**, *Agricultural economics and marketing in the tropics*, Longman, London, 1979. (An excellent non-technical introduction equally relevant to temperate farming systems.)
Barnard, C. S. and **Nix, J. S.**, *Farm planning and control* (2nd edn) Cambridge University Press, Cambridge, 1979.

Chapter 6

Butterworth, B., Davidson, J., Sturgess, I. and **Wiseman, A. J. L.**, *Arable management 1980*, Northwood Books, London, 1980.
Culpin, C., *Farm machinery* (10th edn) Granada, London, 1981.
Culpin, C., *Profitable farm mechanization* (3rd edn) Granada, London, 1975.
Lockhart, J. A. R. and **Wiseman, A. J. L.**, *Introduction to crop husbandry* (4th edn) Pergamon, Oxford, 1978.

Chapter 7

Holmes, W., **(ed.)**, *Grass, its production and utilization*, Blackwell Scientific Publications, Oxford, 1980.

Chapter 8

Castle, M. E. and **Watkins, P.**, *Modern milk production*, Faber & Faber Ltd., London, 1979.
Russell, K., rev. Slater, K., *The principles of dairy farming* (8th rev. edn) The Farming Press, Ipswich, 1980.

Chapter 9

Centre for Agricultural Stategy, *Strategy for the UK forestry industry*, Report No. 6, Reading, 1980.
Nature Conservancy Council, *Nature conservation and agriculture*, London, 1977.
Shoard, M., *The theft of the countryside*, Maurice Temple Smith, London, 1980. (A deliberately provocative presentation of the conservation case.)
Speedy, A. W., *Sheep production: science into practice*, Longman, London, 1980.

Chapter 10

Allen, D. and **Kilkenny, B.**, *Planned beef production*, Granada, London, 1980.
Bateman, D. I. and **Vine, A.**, *Organic farming systems in England and Wales: practice, performance and implications*, University College of Wales, Aberystwyth, 1981.
Mayall, S., (ed.), *Farming organically*, Soil Association, Stowmarket, 1976.
Oelhaf, R. C., *Organic agriculture*, Halsted Press division of John Wiley & Sons, New York, 1978.
United States Department of Agriculture, *Report and recommendations on organic farming*, Washington DC, July, 1980.

Chapter 11

Banks, S., *The complete handbook of poultry keeping*, Ward Lock Ltd., London, 1979.
Buckler, Canon P., *Factory farming – myth and reality*, Arthur Rank Centre, National Agricultural Centre, Stoneleigh, 1980.
Sainsbury, D., *Poultry health and management*, Granada, London, 1980.
Thornton, K., *Practical pig production*, Farming Press Ltd., Ipswich, 1973.
Whittemore, C. T., *Pig production: the scientific and practical principles*, Longman, London, 1980.

Chapter 12

See ref. to **Abbott** and **Makeham** (Ch. 5).
Barker, J. W., *Agricultural marketing*, Oxford University Press, Oxford, 1981.
Bateman, D. I., (ed.), *Marketing management in agriculture*, University College of Wales, Aberystwyth, 1972.
Kohls, R. L. and **Uhl, J. N.**, *The marketing of agricultural products* (5th edn) Collier Macmillan, New York & London, 1980.

Chapter 13

Unilever, *Plant protein foods*, Unilever Educational Booklet, Advanced Series No. 11, Unilever Ltd., London, 1976.

Index